W9-BDY-325

THE MATHEMATICAL CENTURY

THE MATHEMATICAL CENTURY

THE 30 GREATEST PROBLEMS OF THE LAST 100 YEARS

Piergiorgio Odifreddi

Translated by
Arturo Sangalli

With a Foreword by
Freeman Dyson

PRINCETON UNIVERSITY PRESS

PRINCETON AND OXFORD

First published in Italy as *La matematica del Novecento: Dagli insiemi alla complessità*

Published by Princeton University Press,
41 William Street, Princeton, New Jersey 08540
In the United Kingdom: Princeton University Press,
3 Market Place, Woodstock, Oxfordshire OX20 1SY

Odifreddi, Piergiorgio, 1950–
[Matematica del Novecento. English]
The mathematical century : the 30 greatest problems of the last 100 years / Piergiorgio Odifreddi ; translated by Arturo Sangalli.
p. cm.
Includes bibliographical references and index.
ISBN: 0-691-09294-X (acid-free paper)
1. Mathematics—History—20th century. I. Title.
QA26.O3513 2004
510′.9′04—dc21 2003056324

British Library Cataloging-in-Publication Data is available

This book has been composed in Galliard with Helvetica and Futura Condensed Display

Printed on acid-free paper. ∞

www.pupress.princeton.edu

Printed in the United States of America

10 9 8 7 6 5 4 3 2

To Laura,
who frees me from Time and Space
and gives me the joy and peace
forbidden me by Numbers and Points

CONTENTS

FOREWORD

At the beginning of the seventeenth century, two great philosophers, Francis Bacon in England and René Descartes in France, proclaimed the birth of modern science. Each of them described his vision of the future. Their visions were very different. Bacon said, "All depends on keeping the eye steadily fixed on the facts of nature." Descartes said, "I think, therefore I am." According to Bacon, scientists should travel over the earth collecting facts, until the accumulated facts reveal how Nature works. The scientists will then induce from the facts the laws that Nature obeys. According to Descartes, scientists should stay at home and deduce the laws of Nature by pure thought. In order to deduce the laws correctly, the scientists will need only the rules of logic and knowledge of the existence of God. For four hundred years since Bacon and Descartes led the way, science has raced ahead by following both paths simultaneously. Neither Baconian empiricism nor Cartesian dogmatism has the power to elucidate Nature's secrets by itself, but both together have been amazingly successful. For four hundred years, English scientists have tended to be Baconian and French scientists Cartesian.

Faraday and Darwin and Rutherford were Baconians: Pascal and Laplace and Poincaré were Cartesians. Science was greatly enriched by the cross-fertilization of the two contrasting national cultures. Both cultures were always at work in both countries. Newton was at heart a Cartesian, using pure thought as Descartes intended, and using it to demolish the Cartesian dogma of vortices. Marie Curie was at heart a Baconian, boiling tons of crude uranium ore to demolish the dogma of the indestructibility of atoms.

Piergiorgio Odifreddi has done a superb job, telling the story of twentieth-century mathematics in one short and readable volume. My only complaint about this book is that Piergiorgio's account is a little too Cartesian for my taste. It presents the history of mathematics as more orderly and logical than I imagine it. I happen to be a Baconian while Piergiorgio is a Cartesian. We agree about the historical facts. We only disagree about the emphasis. This book is Piergiorgio's version of the truth. My version would be slightly different.

In the Cartesian version of twentieth-century mathematics, there were two decisive events. The first was the International Congress of Mathematicians in Paris in 1900, at which Hilbert gave the keynote address, charting the course of mathematics for the coming century by propounding his famous list of twenty-three outstanding unsolved problems. The second decisive event was the formation of the Bourbaki group of mathematicians in France in the 1930s, dedicated to publishing a series of textbooks that would establish a unifying framework for all of mathematics. The Hilbert problems were enormously successful in guiding mathematical research into fruitful directions. Some of them were solved and some remain unsolved, but almost all of them stimulated the growth of new ideas and new fields of mathematics. The Bourbaki project was equally influential. It changed the style of mathematics for the next fifty years, imposing a logical coherence that did not exist before, and moving the emphasis from concrete examples to abstract generalities. In the Bourbaki scheme of things, mathematics is the abstract structure included in the Bourbaki textbooks. What is not in the textbooks is not mathematics. Concrete examples, since they do not appear in the textbooks, are not mathematics. The Bourbaki program was the extreme expression of the Cartesian style of mathematics. It narrowed the scope of mathematics by excluding all the beautiful flowers that Baconian travelers might collect by the wayside. Fortunately, Piergiorgio is not an extreme Cartesian. He allows many concrete examples to appear in his book. He includes beautiful flowers such as sporadic finite groups and packings of spheres in Euclidean spaces. He even includes examples from applied as well as from pure mathematics. He describes the Fields Medals, which are awarded at the quadrennial International Congress of Mathematicians to people who solve concrete problems as well as to people who create new abstract ideas.

For me, as a Baconian, the main thing missing in this book is the element of surprise. When I look at the history of mathematics, I see a succession of illogical jumps, improbable coincidences, jokes of nature. One of the most profound jokes of nature is the square root of −1 that the physicist Erwin Schrödinger put into his wave equation in 1926. This is barely mentioned in Piergiorgio's discussion of quantum mechanics in his chapter 3, section 4. The Schrödinger equation describes correctly everything we know about the behavior of atoms. It is the basis of all of chemistry and most of physics. And that square root of −1 means that nature works with complex numbers and not with real numbers. This discovery came as a complete surprise, to Schrödinger as well as to everybody else. According to Schrödinger, his fourteen-year-old girlfriend Itha Junger said to him at the time, "Hey, you never even thought when you began that so much sensible stuff would come out of it." All through the nineteenth century, mathematicians from Abel to Riemann and Weierstrass had been creating a magnificent theory of functions of complex variables. They had discovered that the theory of functions became far deeper and more powerful when it was extended from real to complex numbers. But they always thought of complex numbers as an artificial construction, invented by human mathematicians as a useful and elegant abstraction from real life. It never entered their heads that this artificial number system that they had invented was in fact the ground on which atoms move. They never imagined that nature had got there first.

Another joke of nature is the precise linearity of quantum mechanics, the fact that the possible states of any physical object form a linear space. Before quantum-mechanics was invented, classical physics was always nonlinear, and linear models were only approximately valid. After quantum mechanics, nature itself suddenly became linear. This had profound consequences for mathematics. During the nineteenth century, Sophus Lie developed his elaborate theory of continuous groups, intended to clarify the behavior of classical dynamical systems. Lie groups were then of little interest either to mathematicians or to physicists. The nonlinear theory was too complicated for the mathematicians and too obscure for the physicists. Lie died a disappointed man. And then, fifty years later, it turned out that nature was precisely linear, and the theory of linear representations of Lie algebras was the natural language of particle physics. Lie groups and Lie algebras were reborn as one of the central themes of twentieth-century mathematics, as discussed in Piergiorgio's chapter 2, section 12.

A third joke of nature is the existence of quasi-crystals, briefly discussed by Piergiorgio at the beginning of his chapter 3. In the nineteenth century, the study of crystals led to a complete enumeration of possible discrete symmetry-groups in Euclidean space. Theorems were proved, establishing the fact that in three-dimensional space discrete symmetry groups could contain only rotations of order three, four, or six. Then in 1984 quasi-crystals were discovered, real solid objects growing out of liquid metal alloys, showing the symmetry of the icosahedral group which includes fivefold rotations. Meanwhile, the mathematician Roger Penrose had discovered the Penrose tilings of the plane. These are arrangements of parallelograms that cover a plane with pentagonal long-range order. The alloy quasi-crystals are three-dimensional analogs of the two-dimensional Penrose tilings. After these discoveries, mathematicians had to enlarge the theory of crystallographic groups so as to include quasi-crystals. That is a major program of research which is still in progress.

I mention in conclusion one of my favorite Baconian dreams, the possible connection between the theory of one-dimensional quasi-crystals and the theory of the Riemann zeta-function. A one-dimensional quasi-crystal need not have any symmetry. It is defined simply as a nonperiodic arrangement of mass-points on a line whose Fourier transform is also an arrangement of mass-points on a line. Because of the lack of any requirement of symmetry, quasi-crystals have much greater freedom to exist in one dimension than in two or three. Almost nothing is known about the possible abundance of one-dimensional quasi-crystals. Likewise, not much is known about the zeros of the Riemann zeta-function, described by Piergiorgio in his chapter 5, section 2. The Riemann Hypothesis, the statement that all the zeros of the zeta-function with trivial exceptions lie on a certain straight line in the complex plane, was conjectured by Riemann in 1859. To prove it is the most famous unsolved problem in the whole of mathematics. One fact that we know is that, if the Riemann Hypothesis is true, then the zeta-function zeros on the critical line are a quasi-crystal according to the definition. If the Riemann Hypothesis is true, the zeta-function zeros have a Fourier transform consisting of mass-points at logarithms of all powers of prime integers and nowhere else. This suggests a possible approach to the proof of the Riemann Hypothesis. First you make a complete classification of all one-dimensional quasi-crystals, and write down a list of them. Collecting and classifying new species of objects is a quintessentially Baconian activity. Then you look through the list and see whether the zeta-function zeros are there. If the zeta-function zeros are there, then you have proved the Riemann Hypothesis, and you need only wait for the next International Congress of Mathematicians to collect your Fields Medal. Of course the difficult part of this approach is collecting and classifying the quasi-crystals. I leave that as an exercise for the reader.

Freeman Dyson
Institute for Advanced Study
Princeton, New Jersey, USA

ACKNOWLEDGMENTS

I wish to thank John Hubbard and Peter Kahn for the initial inspiration, and Claudio Bartocci, Cinzia Bonotto, Umberto Bottazzini, Lionello Cantoni, Alberto Collino, Vittorio De Alfaro, Simonetta Di Sieno, Michele Emmer, Livia Giacardi, Gabriele Lolli, Cristina Mataloni, Andrea Moro, Alessandro Panconesi, Tullio Regge, and Paolo Valabrega for their help during the production and the final corrections.

THE MATHEMATICAL CENTURY

INTRODUCTION

The world described by the natural and the physical sciences is a concrete and perceptible one: in the first approximation through the senses, and in the second approximation through their various extensions provided by technology. The world described by mathematics is instead an abstract world, made up of ideas that can only be perceived through the mind's eyes. With time and practice, abstract concepts such as numbers and points have nevertheless acquired enough objectivity to allow even an ordinary person to picture them in an essentially concrete way, as though they belonged to a world of objects as concrete as those of the physical world.

Modern science has nonetheless undermined the naive vision of the external world. Scientific research has extended its reach to the vastness of the cosmos as well as to the infinitesimally small domain of the particles, making a direct sensorial perception of galaxies and atoms impossible—or possible only indirectly, through technological means—and thus reducing them in effect to mathematical representations. Likewise, modern mathematics has also extended its domain of inquiry to the rarefied abstractions of structures and the meticulous analysis of the foundations, freeing itself completely from any possible visualization.

Twentieth-century science and mathematics thus share a common difficulty to explain their achievements in terms of classical concepts. But these difficulties can be overcome: often it is only the superficial and futile abstractions that are difficult to justify, while the profound and fruitful ones are rooted in concrete problems and intuitions. In other words, a good abstraction is never an end in itself, an art-for-art's-sake conception, but it is always a necessity, an art-for-humans creation.

A second difficulty in any attempt to survey twentieth-century science and mathematics is the production explosion. Mathematicians, once a small group that often had to earn their living by means other than their trade, are today legion. They survive by producing research that too often has neither interest nor justification, and the university circles in which the majority of mathematicians work unwisely encourage them to "publish or perish," according to an unfortunate American motto. As a result of all this, there are now hundreds of specialized journals in which year after year hundreds of thousands of theorems are published, the majority of them irrelevant.

A third kind of difficulty is due to the fragmentation of mathematics that began in the 1700s, and which became pathological in the 1900s. The production explosion is one of its causes, but certainly not the only one. Another, perhaps even more significant cause is the very progress of mathematical knowledge. The problems that are simple and easy to solve are few, and once they have been solved a discipline can only grow by tackling complex and difficult problems, requiring the development of specific techniques, and hence of specialization. This is indeed what happened in the twentieth century, which has witnessed a hyperspecialization of mathematics that resulted in a division of the field into subfields of ever narrower and strictly delimited borders.

The majority of these subfields are no more than dry and atrophied twigs, of limited development in both time and space, and which die a natural death. But the branches that are healthy and thriving are still numerous, and their growth has produced a unique situation in the history of mathematics: the extinction of a species of universal mathematicians, that is, of those individuals of an exceptional culture who could thoroughly dominate the entire landscape of the mathematics of their time. The last specimen of such a species appears to have been John von Neumann, who died in 1957.

For all these reasons, it is neither physically possible nor intellectually desirable to provide a complete account of the activities of a discipline that has clearly adopted the typical features of the prevailing industrial society, in which the overproduction of low-quality goods at low cost often takes place by inertia, according to mechanisms that pollute and saturate, and which are harmful for the environment and the consumer.

The main problem with any exposition of twentieth-century mathematics is, therefore, as in the parable, to separate the wheat from the chaff, burning up the latter and storing the former away in the barn. The criteria that might guide us in a selection of results are numerous and not at all unambiguous: the historical interest of the problem, the seminal or final nature of a result, the intrinsic beauty of the proposition or of the techniques employed, the novelty or the difficulty of the proof, the mathematical consequences or the practical usefulness of the applications, the potential philosophical implications, and so on.

The choice we propose to the reader can only be a subjective one, with both its positive and its negative aspects. On the one hand, this choice must be made within the bounds of a personal knowledge that is inevitably restricted from a general point of view. And on the other hand, the choice must result from a selection dictated by the author's particular preferences and taste.

The subjective aspects of our choice can nevertheless be minimized by trying to conform to criteria that are in some sense "objective." In the present case, our task has been facilitated by two complementary factors that have marked the development of mathematics throughout the century. Both are related, as we shall see, to the International Congresses of Mathematicians. As in the case of the Olympic Games, these meetings take place every four years, and those invited to present their work are the ones whom the mathematical community considers to be its most distinguished representatives.

The first official congress took place in Zurich in 1897, and the opening address was given by Henri Poincaré, who devoted it to the connections between mathematics and physics. Paris hosted the second congress in 1900, and this time David Hilbert was chosen to open the meeting. The numerological factor prevailed over his desire to reply, three years later in time, to Poincaré's speech, and Hilbert chose rather to "indicate probable directions for mathematics in the new century."

In his inspired address, he gave, first of all, certain implicit clues that shall guide our choice of topics: the important results are those that exhibit a historical continuity with the past, bring together different aspects of mathematics, throw a new light on old knowledge, introduce profound simplifications, are not artificially complicated, admit meaningful examples, or are so well understood that they can be explained to the person in the street.

But Hilbert's address became famous above all for his explicit list of twenty-three open problems that he considered crucial for the development of mathematics in the new century. As if to confirm his lucid foresight, many of those problems really turned out to be fruitful and stimulating, especially during the first half of the century—and we shall examine some of these in detail. In the second half of the century, the thrust from Hilbert's problems petered out, and mathematics often followed paths that did not even exist at the beginning of the century.

To guide us during this period it is useful to turn our attention to a prize created in 1936 and awarded at the Interna-

Fig. I.1 The Fields Medal.

tional Congress to mathematicians under age forty who have obtained the most important results in the past few years. The age limit is not particularly restrictive, given that most significant results are in fact obtained during a mathematician's youth. As Godfrey Hardy put it in *A Mathematician's Apology*: "No mathematician should ever allow himself to forget that mathematics, more than any other art or science, is a young man's game."

The prize was established in memory of John Charles Fields, the mathematician who came up with the idea and obtained the necessary funds. It consists of a medal bearing an engraving of Archimedes' head and the inscription *Transire suum pectus mundoque potiri*, "to transcend human limitations and to master the universe" (fig. I.1). For this reason the prize is nowadays known as the Fields Medal.

This award is considered the equivalent of the Nobel Prize in mathematics, which does not exist. What does exist is a story, widely circulated in mathematical circles, according to which the absence of a Nobel Prize in mathematics would have been due to Alfred Nobel's intention to prevent the Swedish mathematician Gösta Mittag-Leffler from obtaining it. In fact, the two men hardly knew each other, and the latter was certainly not the lover of the former's wife, as the story goes, since Nobel was not married. The real reason is simply that the five original prizes (physics, chemistry, medicine, literature, and peace) were dedicated to the disciplines in which Nobel had had a lifelong interest, and mathematics was not one of them.

In the twentieth century, forty-two Fields medals were awarded, two in 1936 and the rest between 1950 and 1998. Since the winners include some of the best mathematicians of the second half of the century, and the results for which the medals were granted are among the top mathematical achievements of the time, we shall often come back to the subject.

A complement to the Fields Medal is the Wolf Prize, a kind of Oscar for life achievement in a field established in 1978 by Ricardo Wolf, a Cuban philanthropist of German origin who was ambassador to Israel from 1961 to 1973. As is the case for the Nobel Prize, the Wolf Prize has no age restriction, is awarded in various fields (physics, chemistry, medicine, agriculture, mathematics, and art), is presented by the head of state in the awarding country's capital (the king of Sweden in Stockholm in one case, and the president of Israel in Jerusalem in the other) and involves a substantial sum of money ($100,000, compared to $10,000 for a Fields Medal, and $1 million for a Nobel Prize).

To prevent any misunderstanding, I wish to emphasize that the solutions to Hilbert's problems, and the results for which the Fields Medal or the Wolf Prize were awarded, are only significant landmarks and do not exhaust the landscape of twentieth-century mathematics. It will thus be necessary to go beyond them in order to give as complete an account as possible, within the limits previously established, of the variety and depth of contemporary mathematics.

The decision to focus on the great results which, furthermore, constitute the essence of mathematics, determines the asynchronous character of the book's exposition, which will inevitably take the form of a *collage*. This approach has the advantage of allowing a largely independent reading of the various sections, and the disadvantage of resulting in a loss of unity. This inconvenience could be removed on a second reading, which would allow the reader, having already an overall view of the whole, to revisit the various parts.

CHAPTER 1

THE FOUNDATIONS

Depending on one's own philosophical inclination or personal experience, mathematics may be considered an activity of discovery or one of invention.

In the first case, the abstract concepts with which it deals are thought to possess a true and independent existence in the world of ideas, which is considered to be as real as the physical world of concrete objects. Discovery thus requires a kind of sixth sense allowing us to perceive abstract objects in the same way our five senses perceive concrete ones. The fundamental problem of this perception is clearly its external *truth*, that is, a suitable agreement with the assumed reality.

In the second case, mathematical works are instead measured by the same standards as artistic ones, which deal with imaginary objects such as the characters of a novel or the representations of a painting. Invention thus requires a genuine mathematical talent that permits one to construct imaginary objects in the same manner as the possession of an artistic talent does. The fundamental problem concerning the products of this talent is then their internal *consistency*, the possibility to envision the various parts as an organic whole (in mathematical terms: the lack of contradictions).

Be it discovery or invention, mathematics brings forth objects and concepts that, at first sight, are unusual and unfamiliar. Even today some adjectives reveal the reactions of surprise or uneasiness that greeted certain kinds of numbers when they first appeared: irrational, negative, surd, imaginary, complex, transcendental, ideal, surreal, and so on.

A typical attitude, since the days of the Greeks, has been the attempt to minimize surprise and discomfort as much as possible by setting the mathematical building on solid foundations. The history of mathematics bears witness to successive periods of construction and deconstruction, which inverted the mutual relations among what was considered to be fundamental and which replaced precarious or outdated foundations with others deemed more appropriate.

In the sixth century B.C. the Pythagoreans used the *arithmetic* of the integers and the rational numbers as the foundation of mathematics. The crack that caused the building to collapse was the discovery of geometric magnitudes that could not be expressed as the ratio of two integers, thus showing the unsuitability of rational numbers as a foundation of geometry.

In the third century B.C. the whole building was rebuilt by Euclid on geometrical foundations. The integers and their operations lost their role as primitive objects and were reduced to the measure of segments and their combinations; for instance, the product of two integers was seen as the area of a rectangle.

In the seventeenth century René Descartes introduced a new numerical paradigm, based this time on what is now called *analysis*, that is, on the real numbers. Geometry becomes analytical, and points and geometric entities are reduced to coordinates and equations; for example, lines become first-degree equations.

Two centuries later the circle was fully closed, when mathematical analysis was reduced to *arithmetic*. The real numbers were now defined as sets of their rational approximations, and the essential novelty that allowed the modern mathematicians to take this step was the acceptance of actual infinity, a concept that the Greeks had always rejected.

We shall come back later to all these classical foundations. But the construction and deconstruction process did not stop here: in fact, the twentieth century has seen a thriving of alternatives battling for the favors of the mathematicians, making that century a true period of foundational rebuilding. The essential characteristic of the new foundations is that they are no longer based on the traditional objects of mathematics, such as numbers or geometric entities, but rather on entirely new concepts that have completely changed the identity of the subject, both in form and in substance.

1.1. The 1920s: Sets

When we mentioned the arithmetical foundations of the real numbers we already introduced the key word of twentieth-century mathematics: sets. The fact that sets could be used as the basis for the entire mathematical building was Georg Cantor's great discovery. He was led to it by purely mathematical reasons arising from a study of problems in classical mathematical analysis.

Coming from another direction, in connection with an attempt to show how the concepts and the objects of mathematics were, in their deepest sense, of a purely logical nature, Gottlob Frege had also developed his own approach equivalent to that of Cantor, nowadays known as *naive set theory.*

This theory is based on only two principles that reduce sets to the properties that define them. Foremost, the *extensionality principle*, already stated by Gottfried Wilhelm Leibniz: a set is completely determined by its elements, and two sets with the same elements are therefore equal. And then, the *comprehension principle*: every property determines a set, made up of those objects that satisfy the property; and every set is determined by a property, namely that of being an object of the given set.

The discovery that, on two principles that are so simple and elementary from the logical point of view, it was possible to base the whole of mathematics was considered to be the culmination of its history: geometry had been reduced to mathematical analysis, the latter reduced to arithmetic, and now the work of Cantor and Frege showed that arithmetic was in turn reducible to set theory, that is, to pure logic.

But this was too good to be true, and one of the first discoveries of the twentieth century was precisely that such a simple foundation was inconsistent, hence the "naive" qualifier. In 1902, Bertrand Russell proved that the comprehension principle was contradictory, by providing an argument known as the *Russell paradox.*

Basically, sets of objects are divided into two types, depending on whether or not the set is one of the objects contained in the set itself; in other words, depending on whether or not the set belongs to itself. For instance, the set of all sets with more than one element belongs to itself, for it clearly has more than one element; and the set of all sets with exactly one element does not belong to itself, because it also has more than one element.

The question is now: Does the set, S, say, of all sets that do not belong to themselves belong to itself? If the answer is yes, then S cannot be in the collection of the sets that do not belong to themselves, so that S does not belong to S, that is, to itself. If the answer is no, then S is one of those sets that do not belong to themselves, and therefore it must, by definition, be an element of S, so that S belongs to itself.

The solution or, better, the removal of Russell's paradox requires restricting the comprehension principle and making the distinction between *sets* and *classes.* A set is simply a class that is an element of some other class. All sets are therefore classes, but not all classes are sets, and those that are not are called *proper classes.*

If we attempt to apply Russell's argument to the class of all sets, C, say, that do not belong to themselves, we are up for a surprise. In fact, C cannot belong to itself, for then it would be a set that does not belong to itself. Hence, C does not belong to itself, and therefore either C is not a set, or it belongs to itself. Since this latter possibility has just been excluded, the former must hold. In other words, what we have found is not a paradox but a proof of the fact that the class of all sets that do not belong to themselves is a proper class.

Surely, the class of all *classes* that do not belong to themselves leads to a contradiction, just as before. The comprehension principle is therefore reformulated by stipulating that a property of *sets* always determines a class. But in this new form the principle loses much of its strength, because it only allows one to define classes out of sets, which already have to be defined in some way or another.

There are no painless or elegant solutions to the problem, given that the natural one involving the comprehension axiom is inapplicable. It is thus necessary to abandon the analytical, or top-down, approach, and to adopt a synthetic, or bottom-up, one. This is achieved by listing a series of existence principles and construction rules for sets, from which it would be possible to construct what is useful, that is, all the sets that are needed in practice, but at the same time to avoid what is harmful, in other words, the sets that would give rise to paradoxes.

A first list of axioms was proposed by Ernst Zermelo in 1908. His list requires first of all the existence of at least one set, a fact that cannot be established on the sole basis of the comprehension axiom for classes. With a starting point at our disposal, we can then build other sets by means of various operations, the feasibility of which is guaranteed by the axioms. These operations are the set-theoretical counterparts of the arithmetical operations: for instance, taking unions, Cartesian products, and power-sets of sets correspond to forming sums and products of numbers and raising them to a power.

However, all the above operations are not sufficient to guarantee the existence of infinite sets, which are a necessary tool for the reduction of analysis to arithmetic, that is, for the reduction of the real numbers to (infinite) sets of integers. A further axiom thus asserts the existence of an *infinite set*, for instance one whose elements satisfy all the remaining axioms of Zermelo's theory, and which therefore contains in particular all the successive powers of a finite set.

Zermelo's list was updated in 1921 by Abraham Fraenkel, by the addition of an axiom asserting that the values of a function defined on a set also form a set. The resulting axiom system is therefore referred to as *Zermelo-Fraenkel set theory.*

This theory appears to be sufficient for the everyday needs of mathematicians, but this does not mean it will always be so. For example, in the 1960s the work of Alexandre Grothendieck, to which we shall refer later, required the addition of a further axiom: the existence of an *inaccessible set*, whose elements satisfy all the axioms of Zermelo-Fraenkel set theory, and which therefore contains in particular all the successive powers of an infinite set.

More generally, in the second half of the century new axioms were added, guaranteeing the existence of ever larger sets known as *large cardinals*, and the interesting fact is that those axioms allow mathematicians to prove certain results about the integers that cannot be proved without them. In other words, just as in physics there appears to be a connection between the cosmological theory of the universe on a large scale, and the quantum theory of the universe on a small scale, in mathematics there is a link between the global theory of sets and the local theory of numbers.

At any rate, on the basis of Gödel's incompleteness theorem, to which we shall come back, it is impossible to formulate a complete axiom system for set theory, or even only for number theory. Thus, any given extension of the Zermelo-Fraenkel system is bound to be provisional, and to be replaced by subsequent extensions made necessary by a continually improving, but never final, understanding of the notion of a set.

1.2. The 1940s: Structures

Set theory was the nineteenth-century culmination of the reductionist conception of mathematics, which through logical analysis reduces geometry to analysis, analysis to arithmetic, and arithmetic to logic. But the logical analysis of mathematics suffers from the same limitations as literary criticism—namely, to be of interest to specialists but not to authors or general readers, in this case to appeal to logicians but not to mathematicians.

In the eyes of the professional mathematician, set theory had (and still has) two obvious disadvantages. First of all, just as atomic theory has not affected the macroscopic perception of everyday objects, the reduction of mathematical objects to sets has had no influence on mathematical practice. For example, when we count, we do not think of natural numbers as classes of equipotent sets.

Moreover, if paradoxes have worried logicians, they have largely been ignored by mathematicians, who generally see (in)consistency as a problem not of mathematics itself, but of its formal presentations—in this particular case, a problem of the *theory* of sets, not of its *practice*. Zermelo-Fraenkel theory has therefore been perceived as a complicated solution to an irrelevant problem.

In conclusion, set theory appears to have brought the professional mathematician only two benefits, both essential, but independent of any particular axiomatization. On the one hand, we have a theory of infinite sets, that is, as Hilbert put it, "that paradise created by Cantor from which no one will be able to chase us away." And on the other hand, we have a convenient language in which to formulate the increasingly abstract concepts produced by modern mathematical practice.

In the 1930s a group of French mathematicians, known under the collective name of Nicolas Bourbaki, set out to establish mathematics in a manner more appealing to mathematicians and found a solution in an analysis that was not logical but structural. The group undertook the infinite—and hence never completed—project of writing a treatise describing the state of the art of contemporary mathematics. Its title, clearly inspired by Euclid, was *Elements of Mathematics*, and the first volume was published in 1939.

Like Euclid's seminal work, Bourbaki's treatise was divided into books, the first six of which were devoted to foundations. Already their titles indicate the reduced foundational role reserved to sets, for only the first book deals with set theory. The remaining five books feature algebra, topology, functions of a real variable, topological vector spaces, and integration.

In 1949 Bourbaki summarized his philosophical approach, by then the prevalent one, in an article with the revealing title: "Foundations of Mathematics for the Working Mathematician" (rather than for the logician). In that article it was claimed that the whole of contemporary mathematics could be built on the notion of a *structure*, and the treatise that was being written was presented as the concrete proof of the claim.

The fundamental idea of the concept of a structure may be explained with an example. In the theory of sets, real numbers are artificially defined as sets of integers, and their operations and relations are artificially reduced to operations and relations among sets. But in Bourbaki's approach the real numbers and their operations and relations are considered given, and their properties are characterized in an abstract manner.

From a first point of view, it is a question of describing the properties of the sum and the product. For instance, there exist two identity elements, 0 for the sum and 1 for the product; both operations are associative and commutative, inverses exist (except for division by 0), and the product distributes over the sum. These properties are embodied in a general study of *algebraic structures*, whose most common examples are monoids, groups, rings, and fields. The real numbers are then an example of a field.

From a second point of view, it is rather a matter of describing the properties of the order relation. For instance, any two real numbers are comparable; between any two given numbers there is always a third one, and there are no gaps. These properties pertain to a general study of *ordered structures* and are expressed with the concepts of total order, density, and completeness.

Finally, from a third point of view, it is not the properties of individual real numbers that are to be described but rather those of their neighborhoods. For example, the real numbers constitute a set without gaps; every pair of real numbers may be separated by open intervals, and infinitely many open intervals are necessary to cover the entire set of reals. These properties arise in a general study of *topological structures*, and they are expressed with the concepts of connectedness, separability, and (the lack of) compactness.

The above three isolated points of view may then be combined. For example, the operations of sum and product are compatible with both the order and the topological structures, in the sense that they preserve them (except for the multiplication by a negative number, which reverses the order). These properties belong to a general study of *ordered algebraic structures* and *topological algebraic structures*, whose operations are compatible, respectively, with the order and the topological structures. The real numbers then provide an example of an ordered field as well as of a topological field.

Structures already existed before Bourbaki, but the significance of his work was to show that they could be used as a foundation for mathematics. This approach was a great success, for a sufficiently small number of *mother structures* proved enough to treat a large number of interesting cases with optimal efficiency. Bourbaki's influence is today apparent in the modern division of mathematics, no longer into the classical arithmetic, algebra, analysis, and geometry, but rather into a large diversity of hybrids, such as topological algebra or algebraic geometry.

But the advantages of Bourbakism were not only pragmatic. From a theoretical perspective it was also progress with respect to the set-theoretic approach. The study of bare sets and functions among them was shelved, and the new focus was on structured sets, together with the functions that preserve the structure, a less artificial and less drastic abstraction, which better captured the essence of mathematical objects.

1.3. The 1960s: Categories

While the set-theoretic and the Bourbakist foundations were considered satisfactory for a good part of mathematics, and still are, in some fields the concepts of set and structure turned out to be too restrictive and needed an extension. For instance, as we have already mentioned, Grothendieck had to introduce an inaccessible set, and therefore he had to consider the class of all sets that satisfy the axioms of Zermelo-Fraenkel set theory. But the need to widen the structures approach was also a consequence of theoretical considerations, and not only of practical ones.

While the process leading from a concrete example, such as the real numbers, to an abstract structure, such as a topological field, does preserve some significant properties of the example, it also eliminates many others. Only in exceptional cases does a structure admit essentially only one example, which the structure then completely describes. But when a structure admits many radically different examples, as is generally the case, by focusing on the common features of its multiple realizations, their individual characteristics get blurred.

One way of capturing the variety of examples consists in reversing the abstraction process by considering the class of all possible examples of a certain type of structure, together with all possible functions that preserve that structure. We then obtain the concept of a *category*, introduced in 1945 by Samuel Eilenberg and Saunders MacLane. Their work was a natural complement to that of Bourbaki's, as the fact that Eilenberg was a member of the group indicates—in fact, he was the only one who was not French in the history of the group (Eilenberg also received the Wolf Prize in 1986).

For the concept of a category to be considered a special case of the concept of a structure, a new effort of abstraction is required, namely, to determine what it is that the examples of categories obtained from the various types of structures have in common. Even if, at first sight, their extreme diversity may suggest that these examples have very little in common, the amazing discovery of Eilenberg and MacLane was that they share something essential: the fact of consisting of a class of sets together with functions that can be composed in an associative manner, and among which there is always at least the identity function.

And no less surprising was the observation that, since functions carry automatically with them the set of their arguments as well as the set of their values, there is no need to mention these sets explicitly. In this way, the approach is freed from any remnants of naive set theory, still present in the notion of a set with structure, and it becomes an alternative and totally self-contained foundation for mathematics, resting no longer upon the notions of set and membership but on those of function and composition.

Given that sets and their functions are a particular example of a category, it is enough to characterize their properties in categorical terms in order to reduce the whole of set theory to category theory. Such a characterization was found by William Lawvere in 1964, and, ironically, it constituted a further step of logical analysis. For just as the whole of nineteenth-century mathematics had been reformulated using set-theoretical concepts, these very notions were now reformulated in categorical terms.

Category theory thus turned out to be a global and unifying foundation for mathematics, containing as particular cases Zermelo-Fraenkel's sets and Bourbaki's structures. This state of affairs prompted an additional abstraction process. In the same way as sets can be related to one another by *functions*, and structures of the same type can be equally related by structure-preserving functions known as *morphisms*, it was also possible to connect categories among them through functions that preserve the categorical properties, the so-called *functors*.

Since sets and their functions, or structures and their morphisms, form categories, one might be tempted to say that categories and their functors constitute the *category of all categories*. There is, however, a problem. From the set-theoretic point of view, many categories are proper classes and cannot therefore be members of other classes, and in particular be the objects of another category.

A first solution consists in restricting our attention to the so-called small categories, that is, those categories that are sets. We then obtain the *category of small categories*, which generalizes the notion of class of all sets. This category contains many interesting examples, but obviously neither the category of sets nor the category of structures is among them.

A second solution is the one proposed by Grothendieck to which we have already referred, and which was introduced in this context: extending the theory of sets by the addition of new axioms that would allow one to consider classes of classes, classes of classes of classes, and so forth. Depending on the strength of these new axioms, one obtains categories whose objects are classes, classes of classes, and so on, but without ever arriving at the category of *all* categories.

A third, and perhaps the most satisfactory, solution is an axiomatization of the notion of category itself. It was proposed by Lawvere in 1966, and in this context it plays a role similar to the Zermelo-Fraenkel axiomatization of the notion of a set. Moreover, when Lawvere's axiomatization is restricted to *discrete categories*, that is, to those categories whose only functions are the identity functions, one obtains an axiomatization of set theory in categorical form.

All these developments support the claim that category theory has played a significant role in establishing a foundation of mathematics for the mathematicians. This was manifest in the title of the classic 1971 book by MacLane: *Categories for the Working Mathematician.*

Which does not mean that categories have nothing to offer to logicians. As an example it is enough to consider *type theory*, introduced by Bertrand Russell in 1908 as a possible solution to the paradoxes, and which is a version of naive set theory based on the axioms of extensionality and comprehension. What is new in Russell's system is the fact that there are many types of sets, and that a property of objects of a given type determines a set of the next type. In 1969 Lawvere formulated a categorical version of type theory and obtained the *theory of topoi.* Within this theory it is possible to develop a logic, which turned out to be equivalent to the *intuitionistic logic*, introduced by the topologist Luitzen Brouwer in 1912, and more general than classical Aristotelian logic.

Starting from considerations in algebraic geometry completely different from those of Lawvere's, Grothendieck also arrived, in an independent manner, at the theory of topoi. This theory then turns out to be a point of convergence of many fields, and it allowed mathematicians to identify the reason that prevented set theory to serve as a general foundation for mathematics. Simply put: sets form a topos whose logic is classical, and hence is too simple to account for the complexity of, for instance, topology and algebraic geometry.

1.4. The 1980s: Functions

Set theory provided the logicians with an adequate foundation against paradoxes. For their part, the mathematicians, whose daily work is not in the least affected by the problems arising from the paradoxes, found in the Bourbakist structures and in category theory a foundational framework more in tune with their practice.

But none of the three approaches is satisfactory from the point of view of the computer scientists, who make an extensive use of algorithms and programs that operate on data, that is to say, functions that are applied to arguments. Only category theory deals directly with functions, which are not applied to arguments but are composed among themselves. Thus, theoretical computer science needs an alternative foundation, which it found in the *lambda calculus* proposed by Alonzo Church in 1933.

Church's idea was to attempt a different approach to the foundations of mathematics, parallel to the theory of Cantor and Frege, but based on the concept of a function rather than a set. In his scheme, a function corresponds to a set; an argument of a function corresponds to an element of a set; the application of a function to an argument corresponds to the membership of an element in a set; and the definition of a function by a description of its values corresponds to the definition of a set by a property of its elements.

Naive set theory thus automatically translates into a naive theory of functions. The latter rests upon only two principles, which reduce a function to a description of its values. First of all, the *extensionality principle*: a function is completely determined by its values, and two functions taking on the same values for the same arguments are therefore equal. And second, the *comprehension principle*: every description of values determines a function, and every function is determined by a description of values.

If naive set theory had been able to raise much hope before the discovery of Russell's paradox, the naive theory of functions does not appear nearly as promising. In particular, it is reasonable to expect that the paradox could be easily reproduced in the new context.

In trying to reproduce it we immediately run into a problem, namely, what meaning to attach to negation in the framework of functions. The problem may be set aside momentarily by assuming that there is actually a function n that acts in a way similar to negation. Since Russell's paradox arose by considering the set of those sets that do not belong to themselves, we must now consider the function, f, say, whose value for a given argument is obtained by applying n to the result of applying the argument to itself.

The question is now: What is the result of applying the function f to itself? By the definition just given, such a value is obtained by applying n to the result of the application of the function to itself. Hence, the result of applying f to itself is an argument that remains unchanged by the application of n. This is a contradiction, if we assume that n is a function that changes all the arguments to which it is applied. But nothing forces n to be that way. Rather, the above argument shows precisely that n cannot be that way if the theory is consistent, in the sense that no function may assign different values to the same argument.

Therefore, we have a contradiction only if we know the theory to be inconsistent (in the sense mentioned above). But then the above argument would be useless, because it was precisely that fact (i.e., inconsistency) that it intended to prove.

If, on the contrary, the theory is consistent, the argument shows that none of the functions in the theory can change all its arguments. In other words, every function must leave invariant at least one of its arguments, which for this reason is called a *fixed point*.

Russell's argument is therefore not enough to establish the inconsistency of Church's theory, and this is already a partial result. It is conceivable that some other, more elaborate, argument might succeed, but in 1936 Church and John Barkley Rosser proved a difficult and famous theorem, from which follows that the theory is consistent. A function may not assign any value to an argument, but if it does assign one, this value is unique.

The Church-Rosser theorem also showed that lambda calculus is a peculiar theory, based on naive principles and yet demonstrably consistent, and hence protected from paradoxes, both actual and potential. But the remedy appeared, at first sight, to be worse than the disease. The price to pay for consistency was the impossibility to define within the theory a function analogous to negation, and more generally the impossibility for the theory to contain logic. At a time when the appeal of the program to reduce mathematics to logic was still strong despite the obvious obstacles to its realization, the solution seemed unacceptable, and lambda calculus was not considered a suitable foundation for mathematics.

But already in 1936 Church and Stephen Kleene showed that lambda calculus contained arithmetic. Today, their result may be reformulated as follows: the functions representable in lambda calculus are exactly those that may be described in any one of the usual universal programming languages for a computer. Of course, Church's and Kleene's result was ahead of its time, for computers did not yet exist, and the original formulation of the result could not reveal its full potential. With the advent of computers, this potential became obvious, and the theory was rehabilitated as the proper foundation for computer science.

In particular, the fixed-point theorem became the theoretical justification for self-referential—or recursive—programs, largely used in programming. And the *denotational semantics* for lambda calculus, introduced by Dana Scott in 1969, provided techniques that make it possible to interpret computer programs as true and proper mathematical objects, showing at the same time that computer science may be rightly considered as one of the new branches of modern mathematics. His work earned Scott the Turing Award in 1976, a distinction that is the computer science equivalent of the Fields Medal or the Nobel Prize.

CHAPTER 2 PURE MATHEMATICS

For thousands of years the history of mathematics has been primarily the history of the advances in the understanding of numerical and geometric objects. The last few centuries, on the other hand, and especially the twentieth century, have witnessed the emergence of new and essentially different entities. At first quietly subservient to the study of the classical objects, these entities have later acquired an impulsive independence and inspired what has been termed a new golden age of mathematics.

If, on the one hand, modern mathematics is therefore the product of a development rooted in a body of problems that are classical and concrete, on the other it is also the manifestation of an activity that finds its expression in constructions that are contemporary and abstract. Basically, classical mathematics comprised only four fields, devoted to the study of discrete and continuous objects, that is, numbers and figures. Arithmetic and algebra dealt with the former, geometry and analysis with the latter. But it is not so easy to list the branches of modern mathematics, which are concerned essentially with the study of the various algebraic, topological, and ordered structures, and their combinations.

The dangers of this proliferation, which we have already mentioned in the Introduction, are real. But they are averted at the realization that, despite its apparent fragmentation, the mathematics of the twentieth century exhibits a substantial unity. The archipelago of modern mathematics is in fact linked together by mysterious and invisible underground passages, brought to light by unexpected convergences of results that cause them slowly to emerge.

A symbol of this unity is the episode concerning *Fermat's last theorem*, which we shall discuss at length. Its roots reach deep into the past, back to the study of the natural numbers by the Pythagoreans, which culminated in the third century B.C. with Euclid's *Elements*. In the third century A.D., Diophantus of Alexandria began a study of integer solutions of equations with integer coefficients, which he extensively treated in his *Arithmetic*, a monumental thirteen-volume work of which only six have survived. In the seventeenth century Pierre de Fermat studied Diophantus's treatise and, in the margins of his copy, made forty-eight notes without proof.

By the eighteenth century all of Fermat's observations had been proved except one, which for this reason became known as Fermat's last theorem. This is the fact that, while there exist pairs of squared integers whose sum is also a square (for instance 9 and 16, which add up to 25), there are no pairs of cubes whose sum is a cube, nor two n-powers whose sum is another nth power, for n greater than 2. In the nineteenth century, attempts to prove Fermat's last theorem resulted in great advances in number theory and in the confirmation of the theorem for larger and larger exponents, but they fell short of a general proof.

Such a proof was obtained by Andrew Wiles in 1995, by an indirect approach, at first sight totally unrelated to the problem and through the use of an arsenal of completely abstract techniques. Thus, for the solution of a simple numerical problem whose statement is elementary and classical, it was necessary to resort to a large fragment of modern higher mathematics. The event is symbolic not only of the apparent dynamical, asynchronous, and vertical continuity of a single area of mathematics, but also of the hidden static, synchronous, and horizontal connection among its most disparate areas.

Typical of this vision of mathematics as a unified whole is the *Langlands program.* Put forth in the 1960s by Robert Langlands, it specifies a series of conjectures on the possible connections among diverse areas, and Wiles's proof provides a partial but substantial realization of it. In recognition for this unifying work, Langlands and Wiles were awarded the Wolf Prize in 1995–96.

While it is true that number theory—of which Fermat's theorem is one of the results—is perhaps the discipline in which the connections between synchronism and asynchronism, classicism and modernism, concreteness and abstraction typical of contemporary mathematics manifest themselves in the most spectacular way, it is far from being the only one.

Another symbolic event regards the study of the *circle* and the *sphere*, presumably two of the simplest geometric objects. Archimedes was the first to discover, in 225 B.C., the existence of a mysterious connection between some of their features. The circumference and the area of the circle, as well as the surface area and the volume of the sphere, are in fact all related to the constant π, whose calculation required the development, through several centuries, of methods of various kinds (geometric, algebraic, and analytic).

Despite the apparent simplicity of the circle and the sphere, some significant progress in their study had to wait until the nineteenth century. First of all, the development of sophisticated algebraic and analytic methods was necessary in order to prove that there was no solution to the purely geometric problem of the squaring of the circle (to construct, using only a compass and an [unmarked] ruler, a square whose area equals that of a given circle). Moreover, through the use of topological methods it was possible to distinguish the sphere from the other closed surfaces in three-dimensional space. In short, the sphere is the only surface for which an elastic band stretched over it can be contracted to a single point. Finally, differential methods were used to show that infinitesimal calculus can be extended from the plane to the sphere in a unique way.

Some of the fundamental results of twentieth-century mathematics concern the hypersphere, which is the equivalent in 4-dimensional space of the circle and the sphere in 2- and 3-dimensional space, respectively. One of the most important open problems of modern mathematics that we shall discuss later, the so-called *Poincaré conjecture*, involves the validity of a certain topological characterization of the hypersphere similar to that of the sphere, while it has already been proved that infinitesimal calculus can be extended from the space to the hypersphere in only one way.

Circle, sphere, and hypersphere are particular cases of n-dimensional spheres in $n + 1$-dimensional spaces, and some of the deepest and most important results of modern mathematics, to which we shall return, have been obtained by considering spheres in higher dimensions. For instance, a version of the Poincaré conjecture for spheres of all dimensions greater than 3 has been proved, and many nonequivalent ways of extending infinitesimal calculus to the 7-dimensional sphere have been found.

These and other results have revealed an apparent paradox: as the number of dimensions increases, even if the objects become more difficult to visualize, their mathematical study becomes easier, because there is more room to manipulate them. For example, turning inside out a left-hand glove to transform it into a right-hand one is easy in a 4-dimensional space, but difficult in the 3-dimensional space (although not impossible, by a 1959 theorem due to Stephen Smale).

The above impression is confirmed also at an elementary level, as is the case with the calculation of the number of regular "polyhedra." There are five of them in 3-dimensional space (the famous platonic solids), six in 4 dimensions, but only three in spaces of higher dimensions. Ironically, the cases most difficult to study turned out to be precisely those in 3 or 4 dimensions, corresponding to the ordinary space and the space-time in which we live.

The above examples show how even the study of elementary properties of simple objects, such as the integers and the geometric figures, may require the development of sophisticated techniques and abstract areas of mathematics. Since it is precisely from this perspective that one can justify a posteriori both the objects and the methods of modern mathematics, it is the one we shall adopt in presenting the most significant phases of the discipline.

2.1. Mathematical Analysis: Lebesgue Measure (1902)

By its very definition, geometry (from *geo* "earth," and *metrein* "measure") studies questions concerning lengths of curves, areas of surfaces, and volumes of solids. These problems were systematically tackled beginning with Euclid's *Elements*, which in 300 B.C. provided a geometric foundation for the entire Greek mathematics.

Let us consider, as a specific example, the problem of the area. Euclid never gave a definition of area or of its measure, but he stated certain "common notions" from which the following properties may be deduced: equal "surfaces" have equal areas (*invariance*); the area of a surface obtained by "adding" together a finite number of surfaces equals the sum of these areas (*finite additivity*); a surface contained in another surface has an area that is smaller than or equal to that of the latter (*monotonicity*).

On the basis of the first two of these notions, we can assign an area to any polygon in two steps: first, by assigning an area to every triangle ("base times height divided by 2," for instance); and then by decomposing the polygon into triangles

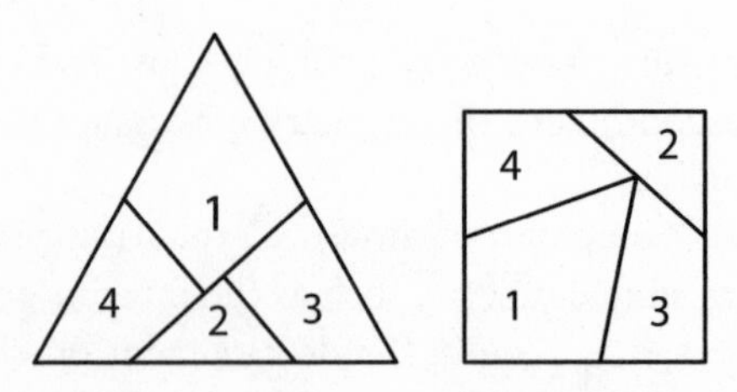

Fig. 2.1 "Squaring" of a triangle.

and adding their areas. Of course, in order for this process to work, one must show that the area of a triangle does not depend on the choice of the base or the height; and also that the area of a polygon is independent of the particular triangulation chosen.

Although these developments were implicit in Euclid's work, his treatment was completely lacking in logical rigor, and in particular it used a number of hidden assumptions that were accurately made explicit only since the nineteenth century. The systematic revision of Euclidean geometry was completed in 1899 with the publication of Hilbert's *Foundations of Geometry.*

In 1833 Jànos Bolyai proved an interesting theorem, a complement to the just mentioned results of Euclid: two polygons with the same area can be decomposed into a finite number of equivalent triangles. In particular, every polygon may be "squared," in the sense that it can first be decomposed into a finite number of triangles, which may subsequently be recomposed so as to form a square with the same area as the polygon. Or, conversely, a square may be transformed into an arbitrary polygon, by recomposing a suitable decomposition of the square into triangles (fig. 2.1).

As for the volumes of polyhedra, it is possible to imagine a similar procedure, in which triangulations are replaced by decompositions into tetrahedra. *Hilbert's third problem* asked whether a result similar to that of Bolyai's was true, that is, whether every polyhedron can be decomposed into a finite number of tetrahedra, which, once recomposed, would form a cube with volume equal to that of the original polyhedron. A negative answer came swiftly, as Max Dehn showed that the construction was already impossible if the polyhedron is simply a tetrahedron.

Once the problem of the area of rectilinear figures such as polygons has been solved, it is a natural step to consider that of curvilinear figures, first and foremost the circle. The idea here is to approximate these figures using polygons, from the inside as well as from the outside. By Euclid's third common notion, the area of the curvilinear figure will fall between the areas of these two approximations, and if the latter tend to a common limit, that limit will also be the area of the curvilinear figure in question.

This general notion is rather recent, having been introduced only in 1887 by Giuseppe Peano, and in 1893 by Camille Jordan. A first special case, using (semi) regular polygons, is Eudoxus's *method of exhaustion*, dating back to the fourth century B.C. and employed around 225 B.C. by Archimedes to calculate the area of the circle and of the surface of the sphere. A second special case, which uses polygons made up of a finite number of rectangles, is the *Riemann integral*, introduced in 1854 by Bernhard Riemann. Using his method we can calculate the area of any surface whose boundary consists of (graphs of) continuous functions.

In fact, from the seventeenth to the nineteenth centuries, the existence of the area of a surface was taken for granted, and the integrals were merely considered as the method for calculating it. This approach was overturned by Augustin Cauchy in 1823 when he defined the area to be the integral itself. This led to the question of which surfaces possessed an area, and in particular which functions had an integral.

The notion of a Riemann integral is a very general one, since it also allows us to calculate the integral of functions having an infinite number of discontinuities, provided these do not form a "nonmeasurable" set. Toward the end of the century, as the examples of functions that were not integrable in Riemann's sense proliferated, it became necessary to be able to define a measure of the set of their discontinuities in order to separate the integrable from the nonintegrable functions.

The notion introduced by Peano and Jordan was not sufficient, and the problem was ultimately solved in 1902 by Henri Lebesgue, with the concept of a *Lebesgue measure*. His idea was essentially to replace Euclid's finite additivity with a countable one: a surface obtained by "adding" together a countable number of surfaces has an area equal to the sum of their areas. Today a surface is considered to have an area (or a solid to have a volume) when this area (or volume) is measurable in Lebesgue's sense.

Armed with his definition of a measurable set, Lebesgue was able to show that a function is integrable in Riemann's sense precisely when the set of its discontinuities has measure zero. This does not prevent the set from being quite large, for instance to possess as many points as the set of all real numbers, although such a set cannot be too "dense."

Moreover, just as the Riemann integral is a special case of Peano's and Jordan's measure, it is possible to define a *Lebesgue integral* as a special case of the Lebesgue measure. All functions that are integrable in Riemann's sense remain so using Lebesgue's definition, and their value is the same; but some Lebesgue integrable functions are not Riemann integrable.

As for the problem of determining which sets are measurable, Giuseppe Vitali showed shortly afterwards that not all sets are. He also discovered that we can do things with the nonmeasurable sets that are impossible with the measurable ones—to the point that, because of our familiarity in dealing with measurable sets, those that are nonmeasurable may appear to possess paradoxical properties.

For instance, in 1914 Felix Hausdorff showed that, given a sphere, it is possible to subdivide its surface into a finite number of pieces (obviously, nonmeasurable ones) in such a way that, once these pieces are rearranged, they form *two* spheres, each one having the same area as the original sphere. And in 1924 Stefan Banach and Alfred Tarski proved a similar result for volumes. In other words, in the space, areas and volumes are not preserved by decomposition into nonmeasurable pieces.

A paradox similar to the above is not possible in the plane. But in 1988 Miklos Laczkovich showed that, in the case of a circle, it is possible to subdivide it into a finite (although very large, around 10^{50}) number of (nonmeasurable) pieces that, rearranged, form a square having the same area. Hence, in the plane, curvature is not preserved by decomposition into nonmeasurable parts.

2.2. Algebra: Steinitz Classification of Fields (1910)

As their very name suggests, the ***natural numbers*** constitute one of the primitive intuitions in mathematics. They are perhaps abstractions of the heartbeat, and so rooted in the notions of time and becoming, in the same way as the geometric points are an abstraction of space and being.

Historically, the first extension of the natural numbers was the introduction of positive rational numbers, which allows the product to have an inverse operation. Since division does not present any major conceptual difficulties, the rationals were well established by the sixth century B.C. and were used by the Pythagoreans as the foundation of their philosophy.

The extension from the natural numbers to the integers, both positive and negative, required instead two essential innovations. The first one was the appearance of zero, without which the problem of the inverse of the sum cannot even be conceived. The zero was introduced in India in the seventh century A.D., and by the Mayas in the second half of the first millennium. The second novelty was the consideration of negative quantities, which are meaningless if numbers are seen as measuring geometric quantities, as the Greeks did. The Indians also introduced the negative numbers in the seventh century A.D., to measure debts.

Combining the two previous extensions, and considering all the rational numbers—positive, negative, and zero—we obtain the first example of a *field*, namely, according to the definition given by Heinrich Weber in 1893, of a set together with operations of sum and product having the usual properties, including invertibility. While the Indians were the first to explicitly adopt the field of rational numbers, the Arabs and later the Europeans rejected negative numbers until the eighteenth century, and still in 1831 Augustus de Morgan denied their reasonableness.

A second typical example of a field is provided by the *real numbers*. The irrationals were discovered by the Pythagoreans and were formally manipulated by the Indians and the Arabs, but they were not considered numbers until the seventeenth century, starting with René Descartes and John Wallis. And it was necessary to wait until the middle of the nineteenth century for a definition of the real numbers, based on the rationals: Richard Dedekind's cuts, in 1858, and the convergent sequences of Georg Cantor (and others) in 1872.

The *complex numbers* were introduced by Gerolamo Cardano in 1545 for the solution of equations of the third degree, and the field operations on them were defined by Raffaele Bombelli in 1572. But in both cases, they only involved formal manipulations on pure symbols, which represented nothing more than "imaginary numbers," and such an attitude persisted until the eighteenth century. The fundamental theorem of algebra, establishing that every polynomial of degree n with complex coefficients possesses n complex zeros, and first proved by Carl Friedrich Gauss in 1799, was the basic result granting the complex numbers an independent existence and provided the first example of an *algebraically closed field*—i.e., a field that contains all the zeros of its polynomials. The formal definition of the complex numbers, as ordered pairs of real numbers, and of their field operations was given by William Hamilton in 1837.

Both Evariste Galois in 1830 and Dedekind in 1871 arrived, with different motivations, to the definition of a whole class of fields through a process of *extension* of the rationals. Given an irrational number a, they considered the smallest set of real (or complex) numbers that forms a field and contains the rationals as well as a. Such a set may be directly generated, starting with a and performing all possible additions, subtractions, multiplications, and divisions (except division by 0). If the number a is the zero of a polynomial with rational coefficients, as in the case of $\sqrt{2}$, the extension is called *algebraic*; otherwise, as in the case of π, it is a *transcendental* extension.

Besides infinite fields, as illustrated by all the preceding examples, there also exist *finite fields.* A simple example is the field of the *integers modulo n*, used to count the hours in a day (with 12 or 24 elements), or the minutes in an hour (with 60 elements). These fields are generated in the same way as the usual integers, starting with 0 and repeatedly adding 1, except that after adding n we find 0 again. For the integers modulo n to form a field it is necessary and sufficient that n be a prime number.

The above examples show how the notions of modern mathematics, that of a field being one of the first significant examples, may serve to unify a great variety of different examples on the basis of some of their common characteristics. However, the very generality of these notions often tends to blur the individual contours of the examples, at the risk of rendering them difficult to grasp. Hence, in order to describe the extensions of these notions, it is imperative to obtain classification results, which are a complementary aspect of abstraction.

One of the first examples of this kind of result was the classification of the possible types of field. It was found by Ernst Steinitz in 1910, based on the notion of *characteristic*. Given a field, we start with the element that acts as 0 and repeatedly add the element that acts as 1. If after p additions we obtain 0 again, p must be a prime number, and we say that the field has characteristic p; if instead we never get back to 0, we say that the field has characteristic 0.

The types of all finite fields can be immediately described by their characteristic. For each prime number p, there are infinitely many finite fields of characteristic p, known as *Galois fields*. The number of elements of each of them is a positive power of p, and for each positive power of p there is exactly one finite field.

Given an arbitrary field, its *prime field* is by definition the set obtained starting with 0 and 1, and performing all possible additions, subtractions, multiplications, and divisions (except division by 0). If the given field has characteristic p, its prime field is a copy of the integers modulo p; if instead the characteristic of the given field is 0, its prime field is a copy of the rational numbers. Now, every field may be obtained by successive extensions starting with its prime field, first a (possibly infinite) series of transcendental extensions, and then a (possibly infinite) series of algebraic extensions.

Conversely, starting with any given field, by a series of algebraic extensions we can construct its *algebraic closure*, namely, the smallest algebraically closed field containing the original field. This illustrates one of the by-products of abstraction, namely the possibility of obtaining general versions of particular results—in this case, the extension to an arbitrary field of the closure process that takes us from the real to the complex numbers, by the adjunction of all possible zeros of polynomials.

2.3. Topology: Brouwer's Fixed-Point Theorem (1910)

In the course of developing set theory as a foundation for mathematics, Cantor often came across some unexpected properties. One of the most surprising of these concerns the geometric notion of dimension: spaces of different dimensions, such as a line and a plane, may in fact have the same number of points and be indistinguishable as sets. This discovery so upset Cantor that he said, after having proved it in 1874: "I see it, but I don't believe it."

Cantor's result certainly did not mean that the notion of dimension was an illusion that would have to be abandoned. Rather, his discovery was an indication that there is a limit beyond which the purely set-theoretic concepts are inadequate and should be replaced by others of a different nature.

In 1910 Luitzen Brouwer showed that *topology*, that is, the study of those properties of geometric objects that are preserved when they are deformed in a continuous manner without breaking them, has the power to distinguish among different dimensions. For example, both a straight line and a plane are made up of a single piece, but a line breaks into two parts if we remove one of its points, while a plane does not (the topological property involved is called *connectedness*). Brouwer proved his theorem of the invariance of dimension in a general setting—more precisely, for any topological object that can be triangulated in a way similar to what can be done for the usual surfaces (such objects are called *complexes*, and the components of the triangulation, *simplexes*).

Brouwer's most significant discovery, however, regards a property of continuous transformations, which are topology's main object of study. Also, in 1910 he proved a *fixed-point theorem* that has become an essential tool in the most diverse domains, from mathematical analysis to economics.

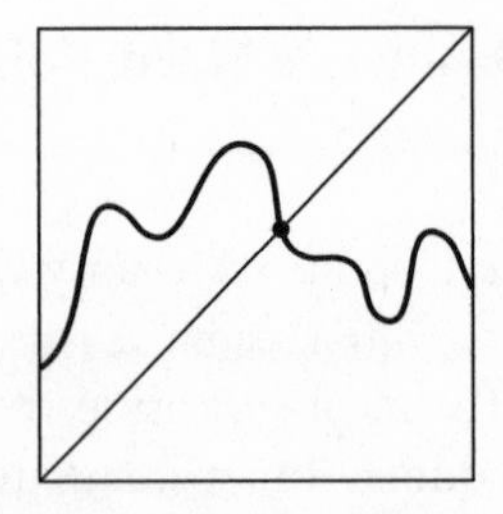

Fig. 2.2 The fixed-point theorem in one dimension.

In the one-dimensional case, Brouwer's theorem amounts to the fact that a continuous function whose arguments and values are all the points of some interval must leave at least one point invariant. This is intuitively obvious, for it means that any curve contained in the unit square, and which extends from one side to the other without gaps, must intersect the diagonal of the square at least once (fig. 2.2).

In the two-dimensional case, Brouwer's theorem says that a continuous function having as arguments and values all the points of a circle must leave invariant at least one point. For instance, if we comb the gravel on a circular flower bed in a continuous fashion, there must be at least one little stone that is not displaced.

The validity of Brouwer's theorem goes beyond what the above two examples may suggest. On the one hand, the theorem applies to the multidimensional counterparts of intervals and circles, such as spheres and hyperspheres. An example of its application in three dimensions is the following: if the wind blows over the entire Earth, it must blow vertically at least at one point, and therefore there must be a cyclone. On the other hand, the theorem holds for all functions defined on simplexes, that is, on surfaces that are sufficiently similar to intervals and circles, in the sense of being bounded and possessing a boundary, and without recesses (these are topological properties known as *compactness* and *convexity*).

The original proof of the theorem established the existence of a fixed point indirectly, without indicating how to find one. Ironically, it was Brouwer himself who later developed a philosophy of mathematics, known as *intuitionism*, which does not accept such nonconstructive proofs. However, in the particular case of the fixed-point theorem, a constructive proof was provided in 1929 by Emmanuel Sperner. With the advent of computers, the necessary calculations became feasible, and today fixed points may be effectively found.

In a different direction, the conditions for the existence of fixed points have been generalized in various ways, and, in particular, some very useful theorems have been proved: in 1922 by Banach, for contractions defined on *complete metric spaces* (these spaces possess, unlike abstract topological spaces, a notion of distance); in 1928 by Knaster and Tarski, for monotonic functions defined on *complete partial orders* (in which every ascending chain of elements has an upper bound); in 1928 by Solomon Lefschetz, for continuous functions defined on *contractable compact complexes*, instead of just on simplexes; and in 1941 by Kakutani, for *semicontinuous functions* whose image sets are all convex, instead of only for continuous functions.

2.4. Number Theory: Gelfand Transcendental Numbers (1929)

Pythagoras's fundamental discovery, in the sixth century B.C., was the existence of a correspondence among music, nature, and mathematics. The harmonic ratios (or intervals) correspond to physical ratios (between the strings of an instrument), and are quantified by numerical ratios (the fractions). The Pythagoreans saw in this agreement not just a coincidence, but the manifestation of a necessity, which they codified in their motto: "All is rational." This should be taken in the literal sense that everything may be described in terms of *rational* numbers (*ratio* meaning precisely "relation").

Their creed was later dealt a deadly blow by their discovery of incommensurable magnitudes, corresponding to *irrational* numbers: in mathematics, the diagonal and the side of a square, whose ratio is $\sqrt{2}$; and in music, the octave and the fifth, whose ratio is $(\log_2 3) - 1$.

Another simple example of an irrational number is $\sqrt[3]{2}$, which solves a problem involving the doubling of an altar. During the 430 B.C. plague in Athens, which killed a quarter of the population, the Athenians consulted the oracle of Apollo at Delos, who asked them to double the volume of the God's cubic altar. The Athenians then built a new altar with the sides twice the sides of the original cube, but in so doing the volume increased eight times, and the plague went on. The correct solution is precisely $\sqrt[3]{2}$, which is the ratio between the sides of two cubes if the volume of one is twice that of the other. Similarly, $\sqrt{2}$ is the ratio of the sides of two squares when the area of one is twice that of the other.

The classification of the real numbers as rational and irrational is rather coarse, and an interesting class of real numbers had already been implicitly introduced by the Greeks: the numbers *constructible* with ruler and compass. For example, $\sqrt{2}$ is constructible, but $\sqrt[3]{2}$ is not. The Greeks suspected the latter fact, but a proof had to wait until 1837, and it was due to Pierre Wantzel. His proof required an algebraic characterization of the numbers constructible with ruler and compass, geometrical operations that correspond to the numerical operations of sum and taking the square root.

All constructible numbers are in any case *algebraic*, that is, they are solutions of algebraic equations with integer coefficients, but the converse is not true. For example, $\sqrt[3]{2}$ is a solution of $x^3 - 2 = 0$, and hence algebraic, but it is not constructible. Numbers that are not algebraic are called *tran-*

scendental, and their existence was first proved in 1844 by Joseph Liouville.

His proof was based on an interesting observation: the fact that an irrational algebraic number cannot be well approximated by rational numbers, in the sense that almost all fractions with denominator q that approximate an irrational solution of a polynomial of degree n have an error of at least $1/q^n$. To construct a transcendental number is therefore sufficient to construct an irrational (i.e., nonperiodic) number that can be well approximated by rational numbers, for example,

$$0.1010010000001000000000000000000000001\ldots,$$

where the first block of zeros after the decimal point has 1 zero, the second block has 1×2 zeros, the third block has $1 \times 2 \times 3$ zeros, and so forth. By truncating the expansion at the 1's after the decimal point, we obtain good rational approximations, correct up to the next 1, which is much farther down the decimal sequence.

In the next one hundred years, many improvements were made to Liouville's observation. The best possible result was obtained by Klaus Roth in 1955: almost all fractions with denominator q that approximate an irrational algebraic number have an error of at least $1/q^{2^+}$, for every real number 2^+ greater than 2 (the result is not valid for 2 itself). For this result Roth was awarded the Fields Medal in 1958.

The best possible extension of the existential part of Liouville's theorem was found by Cantor in 1873: almost all real numbers are transcendental, because there are very few algebraic numbers. More precisely, the latter form a countable set in Cantor's sense, and a set of measure zero in Lebesgue's sense.

But Cantor's result did not say anything about particular real numbers being transcendental or not. As for the number e, the base of the natural logarithms, it was conjectured to be transcendental by Leonhard Euler in 1748 and proved to be so in 1873 by Charles Hermite. From the fact that e is transcendental readily follows that e^2, $\sqrt{e} = e^{1/2}$ and, more generally, e^x for every rational exponent x (different from 0) are also transcendental. In 1882 Ferdinand Lindemann extended the proof and showed that e^x is also transcendental even if x is only algebraic (different from 0), and from this result he derived a large number of consequences. First of all, log x must also be transcendental if x is algebraic (different from 0 and 1), because $e^{\log x} = x$. Moreover, since Euler had shown in 1746 that

$$e^{ix} = \cos x + i \sin x,$$

where i is the imaginary square root of -1 (which, although it is not real, is still algebraic, being the solution of $x^2 + 1 = 0$), it follows that sin x and cos x are both transcendental for x algebraic.

For $x = \pi$ we have a special and particularly important case of Lindemann's result, for then Euler's expression becomes what many consider to be the most beautiful mathematical formula, that is

$$e^{i\pi} = -1.$$

The exponent $i\pi$ thus yields a nontranscendental value of $e^{i\pi}$. It then follows from Lindemann theorem that this exponent cannot be algebraic and, since i is algebraic, π must be a nonalgebraic (i.e., a transcendental) number. The fact that π is transcendental implies, in particular, that π is not constructible, and hence the impossibility of solving another famous Greek problem that remained open for two millennia: the squaring of the circle (with ruler and compass).

By the end of the nineteenth century, not many specific transcendental numbers, apart from π and e, were known, and *Hilbert's seventh problem* asked whether, for example, $2^{\sqrt{2}}$ was transcendental. More generally, Hilbert conjectured that a^b is always transcendental if a is algebraic (different from 0 and 1) and b an irrational algebraic number.

As late as 1919 Hilbert believed that this problem was more difficult to solve than the Riemann hypothesis or Fermat's theorem. In 1929 Alexandr Gelfand showed that e^{π} is transcendental by a new method of proof. This development led in 1930 to the proof by Carl Siegel, who received the Wolf Prize in 1978, that $2^{\sqrt{2}}$ is transcendental, and also in 1934 to the proof of the general Hilbert conjecture by Gelfand and Thorald Schneider. In 1966 Alan Baker brought to a conclusion the results of an entire century by proving that any finite product of transcendental numbers of the types found by Lindemann and/or Gelfand, such as $e^{\pi} 2^{\sqrt{2}}$, is again transcendental. For this result Baker obtained the Fields Medal in 1970.

Despite all these advances, the transcendental numbers remain rather mysterious. The most obvious numbers of which we do not know whether they are transcendental or not are $e + \pi$, $e\pi$ and π^e, but there are many others: for instance, *Euler's constant* γ, which measures the asymptotic difference between the logarithm and the harmonic series—i.e., the (infinite) sum of the reciprocals of the positive integers; and the constant $\zeta(3)$, that is, the sum of the reciprocals of the cubes of the positive integers (the constant $\zeta(2)$, i.e., the sum of the reciprocals of the squares of the positive integers, *is* transcendental, for in 1734 Euler calculated its value to be $\pi^2/6$).

2.5. Logic: Gödel's Incompleteness Theorem (1931)

One of the great mathematical achievements of the nineteenth century was the discovery of *hyperbolic geometry*, that is, of a geometry in which the parallels axiom is false. From the remaining axioms of Euclidean geometry one can show that, given a straight line and a point not in the line, there is *at least* one line parallel to the given line and passing through the given point (the line perpendicular to the perpendicular). The parallels axiom states that there is *only one* such parallel line, and its negation thus implies that there exist *more than one.*

Many mathematicians contributed to the development of hyperbolic geometry, in which the parallels axiom is false, in the hope of showing that such a geometry is inconsistent, and hence proving by contradiction that the parallels axiom is true. In the first half of the nineteenth century, Carl Friedrich Gauss, Nikolai Lobachevsky, and Jànos Bolyai showed in effect that the hypothetical hyperbolic geometry would be very strange. For example, the sum of the angles is not the same for all triangles; a circle does not necessarily pass through three noncollinear points; rectangles do not exist, nor do equidistant lines; and the Pythagoras theorem is false.

However, although strange, hyperbolic geometry did not appear to be contradictory, and in 1868 Eugenio Beltrami showed that it is as consistent as Euclidean geometry: in fact, it is possible to find a model of the hyperbolic plane inside the Euclidean plane. The most famous models of hyperbolic geometry were later found by Felix Klein in 1871 and by Henri Poincaré in 1882. In both cases the hyperbolic plane is a circle without its boundary. In the first model, the lines are Euclidean segments, but the angles must be measured in a different way; in the second model, the angles are Euclidean, but the lines are arcs of a circle perpendicular to the boundary (fig. 2.3).

Once the reduction of the consistency of hyperbolic geometry to that of Euclidean geometry was established, there still remained the question of the consistency of the latter. The Greeks would have required a direct proof, for they had adopted a geometric foundation for all of mathematics, following the discovery by the Pythagoreans of irrational numbers. For example, in the *Elements,* Euclid represented numbers as

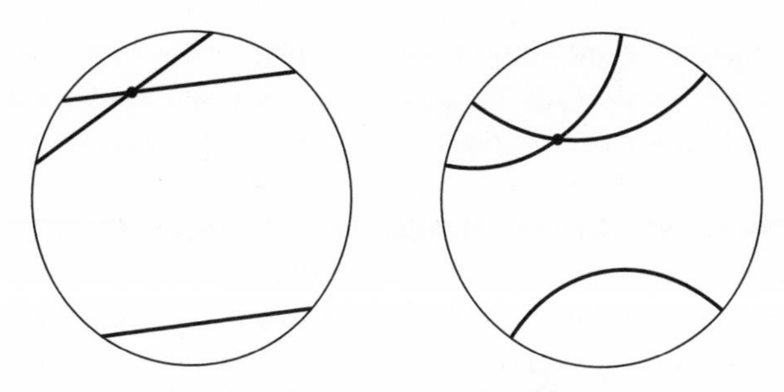

Fig. 2.3 Klein and Poincaré models.

segments, addition as the concatenation of segments, multiplication as the area of a rectangle, and so on.

A reduction in the opposite direction, from mathematics to algebra, was spurred by geography and astronomy. In the second century B.C., Hipparchus, the discoverer of the precession of the equinoxes (the occurrence of the equinoxes earlier in each successive sidereal year) began to use *coordinates* of points to describe given curves, but only with respect to a system chosen each time, depending on the curve. The first to select a *fixed coordinate system* was Oresme, in the fourteenth century. He was still so attached to the geographic usage that he continued to call the coordinates "longitude" and "latitude."

The introduction of a satisfactory algebraic notation, and in particular the use of letters to indicate variables, allowed Pierre de Fermat in 1629, and René Descartes in 1637, to develop *analytic geometry*. The crucial observation was that, by associating points with numbers, one also obtained an induced correspondence between the properties of points and numbers. For example, first- and second-degree equations describe, respectively, straight lines and conics (ellipses, hyperbolas, parabolas).

For both Fermat and Descartes, algebra was in any case subordinated to geometry, and Newton himself in his *Principia* continued to treat the planetary orbits in the geometric fashion of the Greeks rather than in an algebraic way. The change in perspective came with the work of John Wallis, who in 1657 rewrote in an algebraic language two of Euclid's books as well as Apollonius's treatise on conics.

An actual reduction of geometry to algebra only took place in 1899, with Hilbert's *Foundations of Geometry.* He defined an algebraic model of Euclidean geometry as it is currently done today: a point in the plane is an ordered pair of real numbers; a straight line is the set of solutions of a first-degree equation; the distance between two points is defined using Pythagoras's theorem; and congruence is defined through the concept of an isometry (i.e., a distance-preserving linear transformation). But it is not only a question of definitions, for it is necessary to show that an isometry preserves angles as well as distances, and the proof of this fact is not trivial.

By the end of the nineteenth century, the consistence of this geometry had been reduced to that of the theory of real numbers. This game of pass-the-buck could still go on, for instance, by reducing the theory of real numbers to that of integers. In fact, this had actually been done some decades earlier by Karl Weierstrass, Georg Cantor, and Richard Dedekind, which prompted Leopold Kronecker to declare: "God created the integers, all the rest is the work of man." But sooner or later it would have been necessary to prove the consistency of some theory directly, and by methods so elementary that *their own* consistency could not be questioned.

And so, at the dawn of the twentieth century, *Hilbert's second problem* asked for a direct proof of the consistency of the theory of numbers, the reals or the integers. A totally unexpected solution was found in 1931 by Kurt Gödel, who showed that the consistency of any theory containing the theory of the integers cannot be proved within the theory itself. In other words, no theory pretending to be a foundation for mathematics can justify itself and must necessarily search for a justification in some external system. In particular, no consistent theory (containing the theory of the integers) can at the same time be complete, in the sense that all mathematical truths expressible in its language can be proved in the theory—moreover, one of the truths that cannot be proved is precisely its own consistency. For this reason, Gödel's result was called the *incompleteness theorem.*

The impossibility of proving the consistency of a theory from within the theory itself does not preclude the existence of external but nevertheless convincing proofs and does not therefore close the books on Hilbert's second problem. In particular, a significant although obviously not elementary consistency proof of the theory of the integers was given in 1936 by Gerhard Gentzen. His result is the starting point of *proof theory,* whose goal is the search for similar consistency proofs for stronger theories.

2.6. The Calculus of Variations: Douglas's Minimal Surfaces (1931)

According to the *Aeneid* (I, 360–368), at the origin of the foundation of Carthage lies the solution of a mathematical problem. After fleeing Tyre, Queen Dido landed on the North African coast and obtained permission from the local king to choose a piece of land that could be contained within the skin of an ox. The queen cut the skin into a very fine string and used it to enclose an area as large as possible. She chose a semicircular piece of land bordering on the sea, so she would have to enclose only part of the perimeter with the string. Dido had intuitively realized that a circle is the figure that, for a given perimeter, has maximum area. The first mathematical proof of this fact was given by Jacob Steiner in 1838 and completed by Weierstrass in 1872.

Problems of this kind are called maximum or minimum

problems, and in simple cases they can be easily solved by the methods of infinitesimal calculus—more precisely, by expressing them as a function and searching for its critical points, where the derivatives vanish. More difficult cases require the more sophisticated techniques that form the *calculus of variations.* The name derives from the fact that it is the entire function (δf) that varies, and not just an infinitesimal part of it (df).

The first truly variational problem was proposed by Galileo in 1630: given two points that are not vertically one over the other, find the curve (known as the *brachistochrone*) allowing a mass to go from one point to the other in the minimum possible time. Galileo's solution, an arc of circle, was wrong. The problem was proposed again by Jean Bernoulli in 1696, and correctly solved by Newton and Leibniz besides the Bernoulli brothers. The solution is an arc of cycloid, the curve described by a point of a circumference while the latter rolls on a straight line.

Earlier, in the second book of the *Principia* (Scholium to Proposition 34), Newton had already found the first correct solution to a variational problem: find the surface of revolution which, as it moves in water with constant velocity in the direction of its axis, offers the minimal resistance to the motion. He foresaw that the solution to that problem could be useful in shipbuilding, and similar problems later became common in the construction of airplanes and submarines.

The first fundamental result in the calculus of variation is due to Euler. In 1736 he discovered the differential equation that still today lies at the root of the subject, and which establishes a necessary condition for the solution of a variational problem (in the same way as the vanishing of the derivatives is a necessary condition for the solution of maximum or minimum problems). In 1744 Euler wrote a whole book, the first systematic treatment of the subject.

In the first century A.D., Heron of Alexandria had already stated the principle that light travels along paths that minimize both time and space. And Leonardo da Vinci had expressed, in the fifteenth century, the conviction that nature is "economical." These intuitions were extended and systematized in 1744 by Pierre Louis de Maupertuis, in his *principle of minimal action*: natural phenomena occur so as to minimize the action, that is, the product mvs, of mass, velocity, and distance.

Even if Maupertuis's notion of action was not completely satisfactory, its formulation lends mathematical form to the philosophical intuition that at the root of nature's behavior lies a variational principle. Euler had foreseen the possibility of deriving the laws of physics from such a principle, but the first to do it was Lagrange in 1761. He correctly defined action as

$$\int mvds \text{ or } \int mv^2\, dt,$$

and deduced a version of Newton's second law of motion from the principle of minimal action. The final formulation of mechanics in this form is due to William Hamilton, who obtained in 1835 the classical differential equations describing position and momentum as a function of the Hamiltonian H that represents energy.

In 1847 the physicist Joseph Plateau noticed that when a closed wire is immersed into soapy water and then drawn out, a soap bubble is formed inside the wire. This bubble is a surface of minimal area with respect to the perimeter defined by the curve. Soap bubbles thus provide an empirical solution to the problem of finding a surface of minimal area when the shape of the wire is very complex and an explicit solution is difficult to find.

Plateau also stated the following natural problem: prove that for any closed curve in space there is a minimal surface having the curve as its boundary. The problem remains ambiguous if one does not specify what is meant by a closed curve, but after 1887 one could adopt Camille Jordan's definition: a curve is a set of points whose coordinates are the images of continuous functions defined on a certain interval, and this is the definition that applies to Plateau's problem.

The solution had to wait almost a century. It was given by Jessie Douglas in 1931, and for this result he obtained the Fields Medal in 1936, the first one ever awarded. Also for their work on minimal surfaces, among other topics, Enrico Bombieri and Ennio de Giorgi were rewarded, respectively, with the Fields Medal in 1974 and the Wolf Prize in 1990. The calculus of variations has thus seen the value of its shares increase since the beginning of the century, when Hilbert believed that the subject had not been given the recognition it deserved and decided to attract attention to it with his *twenty-third problem*, the only one of a general character on his list. Besides, the *twentieth* and the *nineteenth problems* were also concerned with questions in the calculus of variations, more precisely, the existence and the (analytic) type of the solutions of a large class of variational problems (known as regular). The study of these problems led to the development of a vast area of modern mathematical analysis.

But let us go back to Plateau. One of his experiments consisted in immersing twice in soapy water wires arranged in the shape of a cube. In this case, the resulting bubble was, amazingly, a sort of hypercube, an almost cubic central bubble connected to the original cube by flat sheets (fig. 2.4). In general, similar sheets fill the holes of the surfaces of minimal area obtained with soap bubbles. The existence of minimal surfaces with an arbitrary number of holes, and which therefore cannot be obtained with soap bubbles, was proved in 1987 by David Hoffman and William Meeks through the use of computer-generated images produced in 1983 (fig. 2.5).

Fig. 2.4 Hypercubical soap bubble (from Michele Emmer's film *Soap Bubble*, © 2000 Emmer).

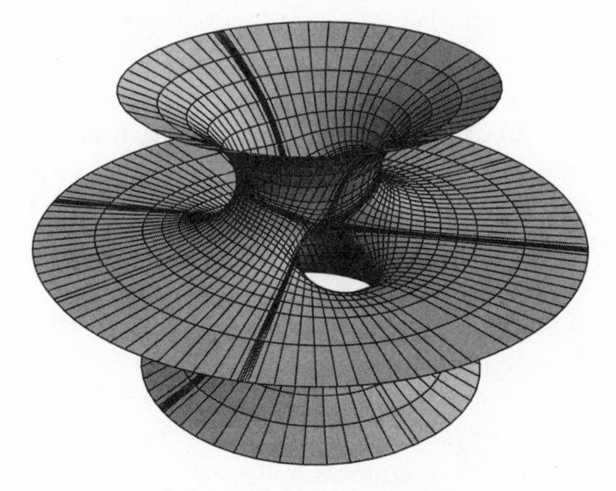

Fig. 2.5 Minimal surface with holes.

2.7. Mathematical Analysis: Schwartz's Theory of Distributions (1945)

It is clear that the Greeks knew some particular curves, such as the conic sections and various kinds of spirals, but they never felt the need to consider the notion of a function in a systematic way. This need arose, however, with the birth of modern science. The study of motion required the consideration of a large class of curves, which naturally included the parabola, the ellipse, and the cycloid; now, these are, respectively, the trajectories of a projectile, of a planet, and of a point of a wheel as it rolls on a plane.

For a long time, the only acceptable way to define functions was by using *formulas*, even though the class of functions was constantly growing as mathematics developed. In the seventeenth century, Descartes demanded that only algebraic equations, that is, polynomials of arbitrary degree in x and y, be used. In the next century, Euler, motivated by the study of the vibrating string, allowed the introduction of analytic expressions involving trigonometric, exponential, and logarithmic functions. He considered these as infinite versions of algebraic functions through their power series expansions. In the nineteenth century, Joseph Fourier, himself motivated by the study of heat, finally also included the trigonometric series.

Fourier's fundamental thesis was that every function can be represented, over an interval, by a trigonometric series. It was precisely in trying to prove this conjecture that Peter Lejeune Dirichlet discovered, in 1829, a famous example of a nonrepresentable function: the function whose value is 1 when the argument is a rational number, and 0 when it is irrational.

This function was not defined by any kind of formula, but a few years later Dirichlet took a bold step: in 1837 he proposed the definition of *function* that is still in use, as a correspondence that to every argument x associates one and only one value y, regardless of how this correspondence is defined.

In a certain sense, the move from definable functions to arbitrary ones is similar to that from algebraic to arbitrary real numbers. In both cases the number of elements increases exponentially, most of which would be impossible to describe given the limitations in the number of possible descriptions. In practice, however, commonly used functions and numbers tend to be explicitly definable in one form or another. Ironically, even Dirichlet's function is no exception, for Peano and René Baire showed that it is analytically representable by the expression

$$f(x) = \lim_{m \to \infty} \lim_{n \to \infty} (\cos m!\pi x)^n.$$

In the course of his study of electromagnetism, Oliver Heaviside introduced in 1893 the improper function δ, defined by the following two properties: its values are all 0, except at the point 0, where the value of the function is infinite; and the region under the graph of the function has area equal to 1. The function δ is patently paradoxical. It differs only at one point from the constant function 0, whose integral is 0, and whatever value we assigned at such a point should not alter the value of the integral. But a single value, undefined and infinity to boot, contributes instead a finite area.

Improper functions such as δ allow mathematicians to express the derivatives of discontinuous functions. For example, δ itself may be considered the derivative of the function H, or *Heaviside function*, which describes an instantaneous unit pulse. Its value is 0 for all arguments less than 0, and 1 otherwise. The justification of the above interpretation involves a limiting process. The function δ is approximated by a function taking on the value 0 almost everywhere, except over an interval around 0; on this interval the value of δ is such that the total area under the graph is equal to 1. As for H, it is approximated by the integrals of the above approximations of δ; all these integrals have the value 0 before the interval and 1 after

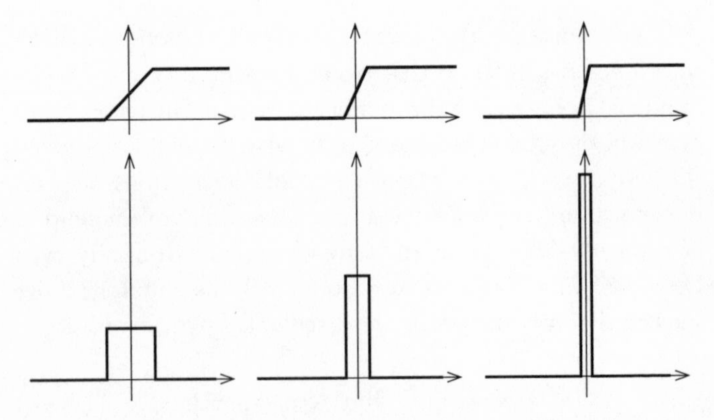

Fig. 2.6 Approximations of the functions H and δ.

it, but over the interval their value varies from 0 to 1 in a continuous manner (fig. 2.6).

The heuristic notions and procedures used by Heaviside were denounced by orthodox mathematicians, and he was even expelled from the Royal Society of London for committing a theoretical sin. As a result, the function δ is nowadays associated not with Heaviside's name but with Paul Dirac's, who used it in his 1930 classical treaty, *The Principles of Quantum Mechanics.* However, Dirac too received his share of severe criticism, in particular from John von Neumann, author of an alternative mathematical formulation of quantum mechanics that we shall discuss in chapter 4. However, thanks to Dirac's reputation, the function δ immediately caught on among physicists, and subsequently also among mathematicians.

An extension of the notion of a function that also included the improper functions was developed by Laurent Schwartz, beginning in 1944 and culminating in 1950 with the publication of the two volumes of his *Theory of Distributions.* In particular, Schwartz developed techniques of differentiation for distributions, showing that every function that is continuous in the classical sense is also differentiable as a distribution. This includes some pathological cases such as the Koch curve, to be discussed later, which does not possess a (classical) derivative at any point. For this work Schwartz was given the Fields Medal in 1950. He later became one of the most famous French intellectuals to take a stand against the Algerian war, and his apartment was blown up by a bomb.

Distributions generalize functions just as real numbers generalize rational numbers, and so classical problems about functions may be extended to distributions. For example, the already mentioned *Hilbert's nineteenth problem* asked which differential operators on functions had only analytic solutions. In 1904 Serge Bernstein came up with the answer: the *elliptic operators*. In his book Schwartz proposed to extend the problem to differential operators on distributions. The solution, due to Lars Hörmander, led to the definition of a new and important class: that of *hypoelliptic operators*, and won the author both the Fields Medal in 1962 and the Wolf Prize in 1988.

One of the fundamental results about elliptic operators is the *index theorem*, proved in 1963 by Michael Atiyah and Isadore Singer. The index of an operator measures the number of its solutions, and it is obtained as the difference between two numbers: that which determines the existence of solutions (i.e., the dimension of the system of linear relations that a solution must satisfy), and the number that determines the uniqueness of the solutions (i.e., the dimension of the space of all solutions). The statement of the theorem establishes that the index is in fact a topological invariant, that is, it does not change by a perturbation of the space on which the operator is defined. This allows for an alternative calculation of the index, and it also lays a fruitful bridge between analysis and topology. The original proof was rather complicated and required the use of the most diverse techniques, from Thom's theory of cobordism (to be examined later) to the K-theory previously developed by Atiyah himself, who for all these achievements was awarded the Fields Medal in 1966. More recently, the index theorem has been reinterpreted in terms of quantum mechanics, and the theory of strings, to which we shall come back, allowed Edward Witten to give a proof that was simpler and easier to understand, and also for this reason to win the Fields Medal in 1990.

2.8. Differential Topology: Milnor's Exotic Structures (1956)

The fact that for a long time the Earth was believed to be flat, and that this appears to be the case when we consider only a small area of it, shows that a spherical surface may be locally Euclidean, even if globally it is not (in technical language, the sphere is locally diffeomorphic to the plane, although not locally isometric to it).

A sphere may therefore be considered a rag ball made up of a large number of very tiny patches that are practically flat, overlapping in a uniform and regular manner. The structure of the sphere as a whole may then be reduced to the structure of the individual patches, on the one hand, and to their position with respect to a canonical reference system, such as the lattice of meridians and parallels, on the other. This way of looking at things allows us to extend the differential calculus to the sphere, that is, to extend the entire arsenal of derivatives and integrals that has been originally created and developed for the Euclidean plane.

In 1854 Bernhard Riemann introduced the notion of a *Riemann manifold* in n dimensions that generalizes the above approach, where extremely small pieces of n-dimensional Euclidean space are patched together in a uniform and regular fashion. We say that such a manifold admits a *differentiable structure* if it is possible to extend to it the usual differential calculus of n-dimensional space, in a similar way as previously indicated in the case of the sphere.

The works of Béla Kerékjártó in 1923, Tibor Rado in 1925, and Edwin Moise in 1952 together led to the proof that all Riemann manifolds of dimension 2 or 3, as well as all Euclidean spaces of dimensions other than 4, admit a unique differentiable structure. It was then believed that the same should be true in general.

However, in 1956 John Milnor showed that the 7-dimensional sphere admits more than one differentiable structure (28, for the record). For this unexpected result, which opened up the new area of differential topology of the so-called *exotic structures*, Milnor received the Fields Medal in 1962 and the Wolf Prize in 1989. In 1969 Michel Kervaire showed instead that there exist manifolds of dimension 10 that do not admit any differentiable structure at all. Taken together with Milnor's result, this proves that neither the existence nor the uniqueness of a differentiable structure are in general guaranteed.

A classification of differentiable manifolds of dimension equal to or greater than 5 was found by Sergei Novikov in 1962, for which he was awarded the Fields Medal in 1970. The most recent developments in differential topology thus concern dimension 4, the only case for which the group of rotations of Euclidean space is not simple (being the product of two copies of the group of rotations of 3-dimensional space). The fundamental results in this domain are due to Michael Freedman and Simon Donaldson, who received for their work the Fields Medal in 1986.

In 1982 Freedman showed how to associate to each simply connected 4-dimensional manifold a symmetric matrix with integers as entries and a determinant equal to ±1, which is defined using the intersection properties of the manifold. Conversely, every matrix of this type corresponds to a manifold. In other words, these matrices define a topological invariant resulting in a classification of simply connected 4-dimensional manifolds. Since already in 1952 Vladimir Rokhlin had shown that not all matrices correspond to differentiable manifolds, Freedman's result establishes the existence of 4-dimensional manifolds that do not admit any differentiable structure.

In a related result, Donaldson showed in 1983 that the only matrices that can arise from differentiable manifolds are unitary matrices. He also found other invariants, which serve to distinguish among differentiable manifolds that are topologically equivalent. In particular, he established the existence of exotic structures on 4-dimensional Euclidean space in which some strange things may happen. For instance, unlike 3-dimensional space, in which any closed and bounded region is contained in a sphere, there are closed and bounded regions of 4-dimensional space that are not contained in any hypersphere. Clifford Taubes and Robert Gompf then showed, in 1985, that the number of exotic structures on 4-dimensional Euclidean space is not only infinite but has the cardinality of the continuum.

An interesting aspect of Donaldson's work is his use of methods in physics to obtain mathematical results. These methods initiated a trend that culminated with Edward Witten's work, which we shall discuss later. Basically, Donaldson replaces Maxwell's equations and the group U(1), which are typical of electromagnetism, with Yang-Mills equations and the group SU(2), which are characteristic of the electroweak theory (to be examined later), and he uses the minimal solutions (known as *instantons*) as geometrical tools. All this suggests the possibility of obtaining other results by a similar use of the same equations but with different groups, for example the group SU(3), typical of chromodynamics.

Going back to differential topology, a question still open is whether the 4-dimensional sphere admits more than one differentiable structure. If the answer is negative, Milnor's theorem on the 7-dimensional sphere would then be the best possible result. Indeed, we already know that the spheres of dimension 2, 3, 5, and 6 have only one differentiable structure. At any rate, the number of differentiable structures on a sphere is strongly linked to the number of dimensions, although it is always finite if the dimension is different from 4. For example, in 8 dimensions, there are 2; in 11 dimensions, 992, in 12 dimensions, only 1; in 15 dimensions, 16,256; in 31 dimensions, more than 16 million.

2.9. Model Theory: Robinson's Hyperreal Numbers (1961)

The first explicit appearance of the *infinitesimals* in mathematics took place in the fifteenth century, when Nicola Cusano defined the circle as a polygon of infinitely many sides of infinitesimal length, and then derived Archimedes' theorem in a few lines: decompose the circle in infinitely many triangles of infinitesimal base and height equal to the radius; since the area of each triangle equals one-half of the base times the height, the area of the circle will then be equal to the circumference (i.e., the sum of the bases of all the triangles) times one-half the radius.

The problem with this approach lies, of course, in the concept of an infinitesimal triangle. If its area is zero, then the circle too should have zero area. If, on the other hand, its area is not zero, then the circle should have infinite area. In neither case would we obtain the right result.

In 1629 Pierre de Fermat employed infinitesimals in the definition of the derivative, which he had introduced as (the measure of the slope of) the tangent to a curve at a given point. He considered a secant passing through two points of the curve: the given point and a second point at an infinitesimal distance h from the first one. He then calculated the (trigonometric) tangent of the (geometric) tangent as the quotient of the incremental changes, just as it is done today. For example, in the case of a parabola:

$$\frac{(x+h)^2 - x^2}{h} = \frac{2xh + h^2}{h} = 2x + h = 2x$$

Here h is assumed to be different from 0 when it is canceled out as a divisor but equal to 0 when it is eliminated at the end, a procedure that could not fail to rise serious doubts as to its consistency.

In 1635 Bonaventura Cavalieri used infinitesimals in the definition of an integral, which he had introduced to calculate areas and volumes. In Cusano's wake, Cavalieri regarded geometric figures as consisting of infinitely many *indivisibles*: the curves made up of points, like "the pearls of a necklace"; the surfaces, of parallel segments, like "the threads of a cloth"; and the solids, of parallel surfaces, like "the pages of a book." But unlike pearls, threads, and pages, the dimensions of these indivisibles were again infinitesimal.

If Leibniz and Newton fully developed the ideas introduced by Fermat and Cavalieri, creating a truly novel methodology for the solution of problems in mathematics and physics, they did not do enough in response to the objections raised by their use of "ghosts of deceased quantities," as a pitiless and critical Bishop Berkeley called them.

In particular, Leibniz based the entire calculus on the notion of an infinitesimal, which he saw as a vanishing, but not vanished, quantity (today we would simply say a non-Archimedean quantity), that is, smaller than any fraction $1/n$, but not equal to zero. And traces of his approach remain even today, in the name *infinitesimal calculus* given to the new discipline, as well as in the notations he invented for derivatives and integrals:

$$\frac{df(x)}{dx} \quad \text{and} \quad \int f(x)\,dx.$$

The derivative is thus represented as the ratio between two infinitesimals (*d* is the first letter of "difference"), and the integral as a sum of indivisibles of infinitesimal length (the symbol ∫ comes from the elongation of the letter S, the beginning of "sum"). The symmetrical use of *d* and ∫ draws attention to the Newton and Leibniz fundamental theorem, according to which taking the derivative and integrating are inverse operations, just as are difference and sum.

Leibniz's approach to calculus through the use of infinitesimals reflected his main concern, which was philosophical, and related to the ultimate constituents of reality (the monads). Newton's approach reflected instead the fundamental applications he had in mind, which were physical, and related to the measure of change (velocity). Unlike Cavalieri, Newton envisaged the geometric figures as generated by continuous motion: curves, from the motion of points; surfaces, from segments; and solids, from surfaces. The derivative was for him not the static ratio between two infinitesimals, but the dynamical "fluxion" of a "flowing" quantity. In his *Principia*, he explicitly declared: "Ultimate ratios in which quantities vanish are not, strictly speaking, ratios of ultimate quantities, but limits to which the ratios of these quantities, decreasing without limit, approach."

This idea was taken up again in 1821 by Augustin Cauchy, who used the concept of limit as a foundation for the whole of calculus. In his formulation, which is the one adopted today, Fermat's example becomes

$$\lim_{h\to 0} \frac{(x+h)^2 - x^2}{h} = \lim_{h\to 0} \frac{2xh + h^2}{h} = \lim_{h\to 0} (2x + h) = 2x.$$

In this way, canceling the number h is justified by the fact that it is a quantity different from 0, while its elimination is replaced by a limit as h tends to 0, without ever having to regard h as being equal to 0. In other words, infinitesimals are variables, not constants.

The precise definition of limit was given by Karl Weierstrass in1859, in the by now familiar "$\varepsilon - \delta$" formulation, and on such grounds the systematic foundation of mathematical analysis could be considered complete. However, this approach had not explained the infinitesimals; it only succeeded in removing them, at the price of a considerable complication of the foundations of calculus.

The rehabilitation of the infinitesimals took place in 1961, when Abraham Robinson showed that using the methods of mathematical logic, in particular the so-called compactness theorem, one can find a class of *hyperreal numbers* that have the same properties as the reals but contain, besides the usual real numbers, also their infinitesimal variants (in a similar way as the real numbers contain, besides the integers, also their decimal variants).

Classical analysis on the real numbers can be extended to a *nonstandard analysis* on the hyperreals, in whose realm Fermat's example becomes absolutely correct: the quantity h is in fact different from 0, so it can be used as a divisor; and even though $2x + h$ and $2x$ are different hyperreal numbers, they have the same real part (just as two decimal numbers may be different but have the same integer part), and are therefore equal from the point of view of the real numbers.

The real numbers may be seen as a completion of the rationals, obtained by going from numbers whose decimal expansion is finite or periodical to numbers whose expansion is infinite. Likewise, the hyperreal numbers may be seen as a completion of the reals, by including numbers whose expansion is doubly infinite.

This suggests the possibility of further completions, with numbers whose expansion is even longer. In 1976 John Conway introduced the *surreal numbers*, whose decimal expansion runs through all the kinds of infinity introduced by Cantor, soon to be examined. We obtain in this way, in a precise sense, the maximum possible completion of the real numbers.

2.10. Set Theory: Cohen's Independence Theorem (1963)

Hilbert's first problem, and hence the one he considered the most important, simply asked how many real numbers there are. From an intuitive point of view the answer to Hilbert's question is obvious: there are infinitely many real numbers.

But Cantor had shown that we cannot simply speak of "infinity" as if this were a well-defined concept. Not only are there many types of infinity, but there are infinitely many types! To make sense of this profusion of infinities, he had rediscovered an abstract approach for comparing the number of elements of any two sets—an approach that Bernhard Bolzano had already used in 1851, and that Duns Scotus in the thirteenth century and Galileo in 1638 had anticipated.

The idea is the following: two sets have the same number of elements if they can be put into a one-to-one correspondence, in other words, if it is possible to pair off the elements of the two sets so that every element of each set has one and only one partner. For example, the class of chairs and the class of people in a room have the same number of elements if no chair is empty and each person is sitting in one chair and does not share his or her seat.

A set has fewer elements than another set if it can be put into a one-to-one correspondence with a subset of the second set, but the second set cannot be put into a one-to-one correspondence with the first. For example, a pair of numbers has fewer elements than a set of three numbers; a set of three numbers has fewer elements than a set of four numbers, and so on. In this way we can easily distinguish between finite sets that have different numbers of elements, as well as between finite sets and infinite ones.

It is, however, natural to expect that all infinite sets would be equivalent, and Cantor's first results did point in that direction. For instance, he showed that the set of all integers may be put into a one-to-one correspondence with the set of the positive integers, by ordering the former as follows:

$$0, 1, -1, 2, -2, 3, -3, \ldots$$

Similarly, as John Farey had already observed in 1816, the (positive) rational numbers may be put into a one-to-one correspondence with the (positive) integers by ordering them according to the sum of numerator and denominator, namely

$$1/1, 1/2, 2/1, 1/3, 2/2, 3/1, 1/4, 2/3, 3/2, 4/1, \ldots$$

(repetitions may be easily eliminated, if one so wishes).

But in 1874 Cantor discovered that the real numbers cannot be put into a one-to-one correspondence with the integers. Every list of real numbers must be incomplete, because it does not contain the real number whose first decimal is different from the first decimal of the first number on the list; its second decimal is different from the second decimal of the second number on the list, and so forth. Thus, there are more real numbers than integers and, with a proof similar to the above, known as the *diagonal method*, Cantor showed in 1891 that for each infinite set, there is another one with more elements than the first.

Since it can be shown that the infinity of the integers is the smallest of all, the infinity of the reals must be greater than the infinity of the integers. A natural question is whether this infinity comes immediately after that of the integers, or if there are other infinities in between—in other words, whether there exist subsets of the real numbers having more elements than the integers but fewer than the set of all reals. In 1883 Cantor conjectured that there are none, and this statement became known as the *continuum hypothesis* ("continuum" is a name given to the set of real numbers).

The first result in connection with this problem was obtained by Kurt Gödel in 1938. Adhering to Wittgenstein's motto, "What we cannot speak about we must pass over in silence," he decided to restrict his attention to the *constructible sets*, the only sets of which we can speak about in a precise hierarchical language. Gödel discovered that the constructible sets form a universe that satisfies all the axioms of Zermelo-Fraenkel set theory as well as the continuum hypothesis. This implies that the negation of the continuum hypothesis cannot be derived from the axioms, unless these are contradictory. In other words, the continuum hypothesis is *consistent* with set theory, in the sense that it cannot be refuted.

A complement to Gödel's result was obtained in 1963 by Paul Cohen. He decided this time to extend the universe to include *generic sets*, which satisfy all the typical properties of sets. Cohen's discovery was that the addition of generic sets to the constructible sets produce universes that satisfy all the axioms of Zermelo-Fraenkel and, in some cases, also the negation of the continuum hypothesis. This means that this negation cannot be derived from the axioms either, unless these are contradictory. Put another way, the continuum hypothesis is *independent* of set theory, in the sense that it cannot be proved nor, as Gödel had already shown, disproved. For his result Cohen received the Fields Medal in 1966.

Hilbert's first problem is hence solved, and the solution is that it cannot be solved with the usual notions of set theory. This does not mean, of course, that some extensions of these notions, which may be considered equally natural, cannot appear in the future and allow mathematicians to decide the continuum hypothesis one way or the other. For the moment, however, it is necessary to keep track of which results in set theory have been proved using the continuum hypothesis (or its negation), and which ones without using it.

2.11. Singularity Theory: Thom's Classification of Catastrophes (1964)

The simplest way to analytically describe curves in the plane is by means of polynomials in x and y, which define the so-called *algebraic curves*. Descartes discovered in 1637 that the polynomials of the first degree describe the straight lines, and those of the second degree the *conic sections* already studied by the Greeks, namely, hyperbola, ellipse, and parabola. Their name derives from the fact that they can all be obtained by projections and sections of a circle, in the sense that projecting the circle from a point one obtains a cone, and intersecting the cone with a plane produces the conic sections.

The third-degree polynomials define the *cubics*, whose study required the use of the new methods of the infinitesimal calculus. Newton discovered in 1676 that there are only eighty types of cubic, and that they can all be obtained by projections and sections of *elliptic curves*, so called because of their role in the calculation of the length of an arc of ellipse (an ellipse, however, is *not* an elliptic curve), and whose general form is

$$y^2 = ax^3 + bx^2 + cx + d.$$

Moreover, there are only five types of elliptic curves, classified according to the possible zeros of the third-degree polynomial to the right of the equal sign (fig. 2.7).

More precisely, four cases are obtained when all three zeros are real numbers. If all are different, the curve consists of two

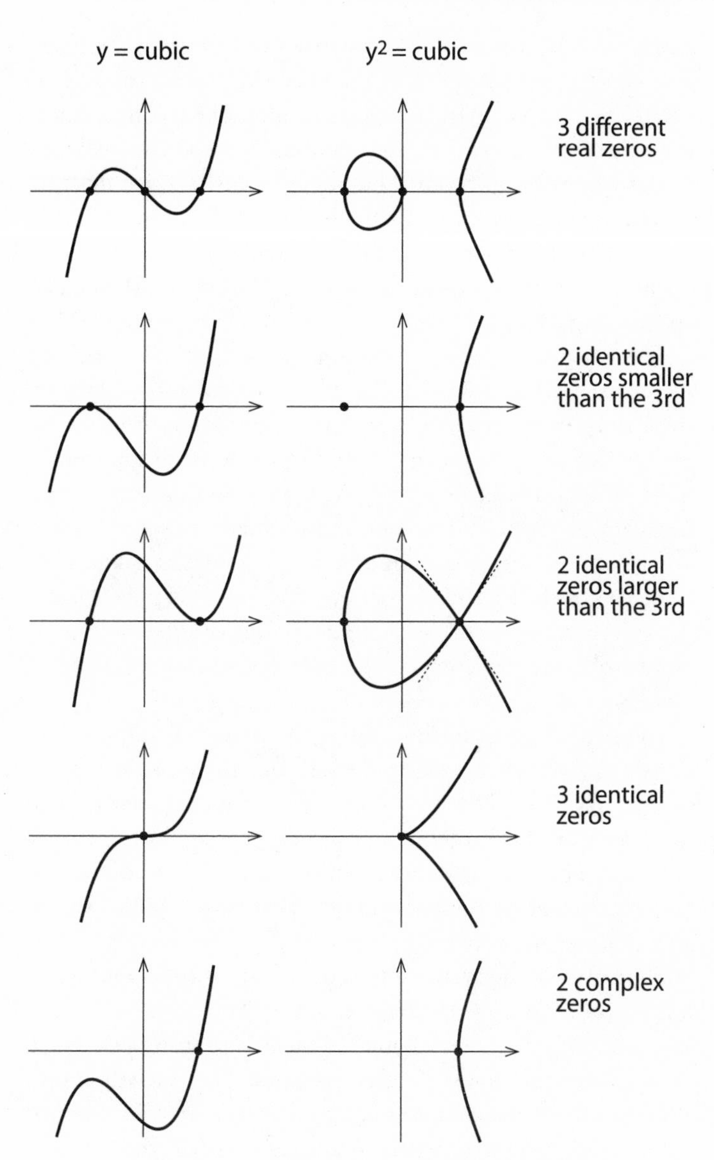

Fig. 2.7 Classification of the elliptic curves.

pieces, one of which is closed. If two of the zeros coincide, they can be smaller or larger than the third zero; in the first case, the curve has an isolated point, and, in the second, a knot. If all three zeros coincide, the curve has a cusp. The fifth case occurs when the polynomial has nonreal zeros, which must be two different complex numbers, because a third-degree polynomial with real coefficients must always have a real zero, and the nonreal zeros always occur in pairs. The curve is then made up of a single smooth piece.

At every point of a conic section the tangent is unique and the curve lies on only one side of it, but for more complex curves this property may not hold, and we are then in the presence of *singular points.* These already occur in some elliptic curves: at a knot or a cusp there are two tangents, in the first case different and in the second identical. At an inflexion point the curve crosses the tangent, and its concavity changes. In 1740 the abbot Jean Paul de Gua de Malves showed that, in general, the singular points of algebraic curves may be obtained by composing knots, cusps, and inflexions in various ways.

The study of nonalgebraic curves is more difficult, and the purpose of *singularity theory* is to deduce the global behavior of the curve from the local knowledge of its singular points. More generally, the theory seeks to classify families of curves or surfaces by reducing them to a small number of types that are determined by their singularities, in a way similar to the above classification of the cubics.

The notion of derivative allowed Fermat, in 1638, and Newton, in 1665, to study the *smooth curves,* that is, the curves that have a derivative at every point, and whose singular points are those where the first derivative vanishes. The smooth curves are reducible, through local deformations, to the smooth curves whose singular points, if any, are regular, that is, with nonzero second derivative. At such points the curve may be approximated by a second-degree monomial—i.e., a parab-

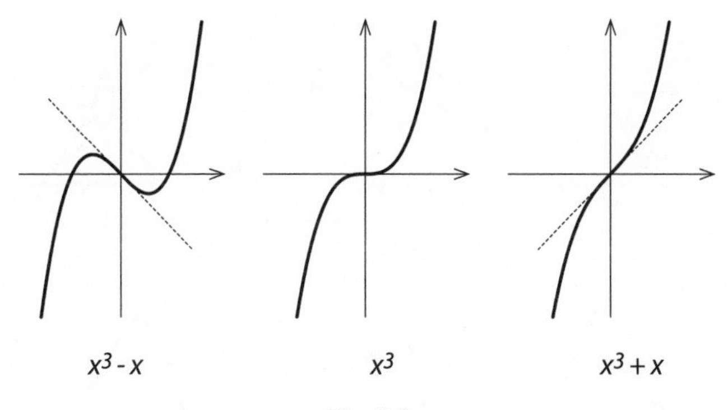

Fig. 2.8

ola—and depending on whether the sign is positive or negative, the parabola opens upwards or downwards, and the singular point is therefore a minimum or a maximum. For example, the cubic x^3 has a singular, nonregular point, that is, an inflexion, at the origin, where the tangent is horizontal. But a small rotation of the tangent suffices to transform the curve into one of type $x^3 + x$, without singularities, or in a curve of type $x^3 - x$, with a maximum and a minimum (fig. 2.8).

An extension of the previous results, from curves to n-dimensional surfaces, was carried out by Marston Morse in 1934. He showed that the *smooth surfaces* are reducible, through local deformations called diffeomorphisms, to smooth surfaces whose singular points, if any, are regular. At such points the surface can be approximated by an algebraic sum of second-degree monomials in each of the variables, that is, by a saddlelike surface. The type of the latter is determined by the number of monomials with positive or negative sign, in other words, by the number of directions in which the saddle opens upwards or downwards.

Morse's theorem completely characterizes the regular singular points, leaving open the problem of characterizing those singular points that are not regular. These are called *catastrophes*, because they correspond to radical bifurcations in the be-

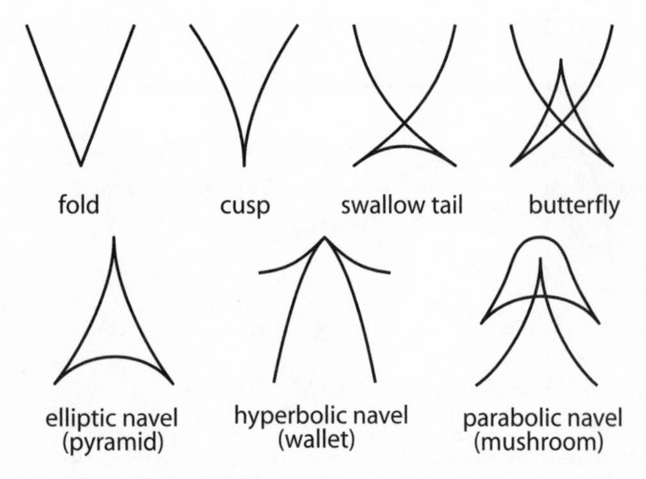

Fig. 2.9 Classification of catastrophes.

havior of the system. The study of surfaces with nonregular singular points is the subject of *catastrophe theory*, which was develolped by René Thom.

For smooth curves, the only catastrophes are the inflexions, at which the curve is flat because it crosses its horizontal tangent. When we consider the case of n-dimensional surfaces there are several possibilities, depending on the number of directions in which the curve is flat, known as the *corank*, and on the minimum number of deformations needed to eliminate the irregularity, or *codimension*. For instance, the cubic x^3 described above has clearly corank equal to 1, and also codimension 1, for it is enough to add to it a single term in order to eliminate the inflexion. Inspired by a 1947 work on cusps due to Hassler Withney, who won the Wolf Prize in 1982, Thom conjectured in 1964 that corank and codimension suffice to classify all catastrophes. More precisely, he believed that in case the codimension is less than or equal to 4, the catastrophes are of only seven types: four of corank 1, that is, folds, cusps, swallow tails, and butterflies; and three of corank 2, that is, pyramids, wallets, and mushrooms (fig. 2.9). For larger codimension, the catastrophes become infinite. Thom's conjecture was proved by John Mather in 1966.

The interest in catastrophe theory stems from the fact that it was one of the first mathematical tools that seemed capable of putting some order into chaos, and of describing regularities in irregular behavior. In his 1972 book *Structural Stability and Morphogenesis*, Thom began to apply catastrophe theory to the study of the most diverse phenomena, from the formation of the embryo to the outbreak of revolutions. These applications of the theory were later driven to extremes by Christopher Zeeman.

From the point of view of the applications, catastrophe theory has nowadays been surpassed in two ways: first, by Ilya Prigogine's theory of dissipative structures and the thermodynamics of irreversible phenomena, for which he was awarded the Nobel Prize in chemistry in 1977; and second, by chaos theory and the dynamics of unstable systems, which we shall treat later.

2.12. Algebra: Gorenstein's Classification of Finite Groups (1972)

It is known since Babylonian times that there is a simple algebraic formula to calculate the solutions of any second-degree equation,

$$ax^2 + bx + c = 0,$$

viz

$$x = \frac{-b \pm \sqrt{b^2 - 4ac}}{2a}.$$

Algebraic formulas to calculate the solutions of any given equation of degree 3 or 4 were found in the sixteenth century by several Italian mathematicians, among them Niccolò Fontana (also known as Tartaglia), Gerolamo Cardano, and Ludovico Ferrari. But Paolo Ruffini in 1799 and Niels Abel in 1824 showed that there are no algebraic formulas to calculate the solutions of an arbitrary equation of the fifth degree.

In 1832 Evariste Galois solved the general problem of determining *which* equations admit solutions that can be expressed by an algebraic formula. To formulate his theory, Galois introduced the concept of a *permutation group* of the solutions. By a permutation of a set, we understand a reordering of its elements. For example, 2 - 3 - 1 is the result of a permutation of 1 - 2 - 3.

In 1849 Auguste Bravais, while studying problems in crystallography, introduced the similar concept of a *symmetry group*. This time one considers geometric transformations that leave a figure invariant with respect to certain criteria, for example, the rotations of a regular polygon in the plane, or of a regular polyhedra in the space. Particularly interesting symmetry groups are those arising from the rotations of the circle or the sphere that are infinite in number (one for each angle of rotation). These groups are examples of *Lie groups*, to which we will come back.

As illustrated by the above examples, groups of different kinds naturally appear in various areas of mathematics, and in 1849 Arthur Cayley introduced the concept of an *abstract group*. This is a set on which an operation is defined among its elements, such that (1) repeated application of the operation to elements of the set always produces an element of the set; (2) there exists a "neutral" element, which plays the same role that 0 and 1 play with respect to the sum and the product, respectively; (3) the operation admits an "inverse," in the same sense as subtraction (or division) is the inverse of the sum (or the product); (4) the operation is associative, as are the sum and the product, viz.,

$$a + (b + c) = (a + b) + c \text{ and } a \times (b \times c) = (a \times b) \times c.$$

The operation need not be commutative, in the sense the sum and the product are, viz.,

$$a + b = b + a \text{ and } a \times b = b \times a,$$

but if it is, the group is called *abelian*.

The generality of the concept of a group makes its application easy but its characterization difficult. An essential simplification, due to Galois, consists in defining the class of *simple groups*, which are the elementary components of groups in the same sense the prime numbers are the basic building blocks of the integers. An operation of factorization for groups can be introduced, and the simple groups are those admitting as factors only themselves and the trivial group (i.e., the group with a single element). The classification problem for groups is therefore reduced to the classification of all simple groups.

The first step was the classification of the *continuous groups* of transformations defined in 1874 by Sophus Lie and called *Lie groups* after him. These may be defined as those groups with a local coordinate system with respect to which the group operations are analytic. The theory of Lie groups, which from the start involves algebra, topology, and analysis, has been and still is a source of deep and difficult problems. One of these, *Hilbert's fifth problem*, asked whether every locally Euclidean group (i.e., admitting a local coordinate system) was a Lie group. It was solved in the affirmative by Andrew Gleason, Deane Montgomery, and Leo Zippin in 1952.

Even if a Lie group has an infinite number of elements, these can be identified by specifiying only a finite number of parameters, known as the *dimension* of the group. For example, the group of rotations of the circle, which is denoted either U(1) or SO(2), has dimension 1 because it is enough to specify the angle of rotation. The group of rotations of the sphere, denoted SO(3), has dimension 3 because it is necessary to specify both the axis of rotation (which can be identified by its latitude and its longitude) and the angle of rotation.

The classification of simple Lie groups was outlined by Wilhelm Killing in 1888 and perfected by Elie Cartan in 1894. To begin with, it was found that there are four infinite families, all made up of groups whose elements are square matrices, that is, with n rows and n columns. The matrices in each family have their own particular properties, for example, SO(n) and SU(n) are, respectively, the groups of special orthogonal matrices and special unitary matrices.[1] In addition, there are five *sporadic groups*, that do not fit in any of the four families but are exceptions, denoted G_2, D_4, E_6, E_7, and E_8, with dimension 14, 52, 78, 133, and 248, respectively.

The theory of Lie groups is today the language that allows physicists to express the unified field theories of *particle physics*. More precisely, it was discovered that the electromagnetic force, the weak nuclear force, and the strong nuclear force preserve certain symmetries: phase rotation symmetries of the fields, symmetries of charge exchange among particles, and symmetries of color exchange among quarks, and that the properties of these symmetries are described by the Lie groups U(1), SU(2), and SU(3). The respective dimensions of these groups are 1, 3, and 8, and correspond to the number of bosons that transmit the three forces: 1 photon, 3 weak bosons, and 8 gluons.

[1] The reason for these names is the fact that the linear transformations defined by unitary matrices preserve the unit of length (i.e., distance), while those defined by orthogonal matrices also preseve orthogonality. In technical terms, a matrix is *special* if its determinant equals 1; *orhtogonal* if the product

The first attempt to describe these symmetries mathematically was due to Chen Ning Yang and Robert Mills in 1954, who used the group SU(2) to describe certain symmetries of the strong interactions (instead of weak ones), providing the first example of what are now called the Yang-Mills equations. The second attempt was the work in 1961 of Murray Gell-Mann, who used the group SU(3) to describe the flavor (instead of the color) symmetries of quarks. In recognition of his achievement he received the Nobel Prize in physics in 1969. Another Nobel Prize in physics was awarded in 1979 to Sheldon Glashow, Abdus Salam, and Steven Weinberg for the identification in 1968 of U(1) × SU(2) as the characteristic group of the electroweak theory. Finally, SU(3) was identified in 1973 by Weinberg, David Gross, and Frank Wilczek as the characteristic group of chromodynamics.

The road toward the final unification of the forces in physics thus goes through the discovery of a suitable Lie group that contains the product U(1) × SU(2) × SU(3). The minimum simple Lie group that satisfies this mathematical condition is SU(5), which has dimension 24, but it does not appear to be suitable from the point of view of physics. The *great unification* based on this group predicts dubious phenomena, such as a too rapid decay of the proton and the existence of magnetic monopoles. Today's most popular candidate for the so-called *theory of everything*, which also includes gravity, is instead a double copy of the largest sporadic group E_8. With a dimension that is twice 248, it predicts the existence of 496 field bosons (incidentally, a perfect number!), of which only the 12 previously mentioned are known.

As for the classification of the finite simple groups, things are more complicated than for the Lie groups. At the end of the nineteenth century, six infinite families were known, as well

---

with its transpose is the identity matrix; and *unitary* if the product with its conjugate transpose is the identity matrix.

as five sporadic finite groups discovered in 1861 by Émile Mathieu in his study of finite geometries, the largest of which had close to 250 million elements.

Of the six families of finite simple groups, four were the equivalents of the families of Lie groups. The fifth family was that of the *cyclic groups*, that is, the already mentioned integers modulo *n*, and the simple cyclic groups are precisely those with a prime number of elements.

The sixth family was that of the *alternating groups* defined by Galois. The starting observation is that every permutation may be obtained by a succession of exchanges of adjacent elements. For instance, the permutation 2 - 3 - 1 may be obtained from 1 - 2 - 3 by first exchanging the first two elements (2 - 1 - 3) and then the last two (2 - 3 - 1). The alternating groups are those whose elements are the permutations obtained by performing an even number of exchanges of adjacent elements, as in the above example, and the alternating groups obtained from permutations on sets with more than four elements are all simple (Galois showed that this is precisely the reason why it is impossible to find general algebraic formulas for the solution of equations of degree greater than 4).

New families were found in 1957 by Claude Chevalley. In particular, every sporadic Lie group gave rise to an entire family of similar groups defined over finite fields. And new sporadic finite groups were discovered in 1965 by Zvonimir Janko. These results opened a new period of discovery that led to the identification of a total of 18 families and 26 sporadic finite groups, the largest of which is a monster of 10^{54} elements. Just as in particle physics, the new groups were often first predicted theoretically, and later "observed in the laboratory." For example, the above-mentioned monster was predicted in 1973 by Bernd Fischer and Robert Griess, and constructed (by hand!) by the latter in 1980.

However, the real problem was to prove that the 18 families and the 26 sporadic groups constituted the desired classification, in the sense that every finite simple group either belongs to one of the families or is one of the sporadic groups. The program for the solution was put forward by Daniel Gorenstein in 1972. The proof, completed in 1985, required the joint efforts of some one hundred mathematicians, is spread over five hundred articles for a total of 15,000 pages, and holds the record for the most complex proof in the history of mathematics.

Gorenstein's program proceeds by cases; it first reduces the possible cases to about one hundred, and then proves for each of them a limited classification theorem. One of the most important cases is that of the simple groups with an odd number of elements. By the *second Burnside conjecture*, dating from 1906, these must be precisely the cyclic groups with a prime number (greater than 2) of elements. The conjecture was proved in 1962 by Walter Feit and John Thompson, in a 250-page article, and for this result Thompson obtained the Fields Medal in 1970 and the Wolf Prize in 1992.

Nevertheless, the classification of the finite groups is not the end of the story. The *first Burnside conjecture*, in 1902, asked whether every group with a finite number of generators (i.e., every element of the group is a combination of them) and which is periodic of order n (i.e., after n combinations with itself every element of the group becomes the identity element) is finite. Since the converse is obvious, the conjecture would have completely characterized the finite groups, but it was disproved in 1968 by Petr Novikov (father of Sergei, Fields medalist in 1970) and S. I. Adian.

A weaker version of the conjecture, already reformulated in the 1930s, asks for the finiteness not of the group in question but only of the number of its finite quotients. This was proved in 1991 by Efim Zelmanov, who obtained for this result the Fields Medal in 1994. He proved the result when n is a prime power, and the general case can be reduced to the special case using the classification theorem for finite groups (a more direct proof of the conjecture is not known).

2.13. Topology: Thurston's Classification of 3-Dimensional Surfaces (1982)

One of the great achievements of nineteenth-century mathematics was the classification of the 2-dimensional surfaces from the topological point of view, that is, by regarding them as rubber sheets that can be deformed at will provided they are not torn apart. From this abstract point of view, an inflated ball and a deflated one are the same surface, even if from the outside one appears to be a sphere and the other a curled up sheet. On the other hand, a ball and a lifesaver are different surfaces, because the ball cannot be deformed into a lifesaver without breaking it.

The classification makes an essential use of the concept of a nonorientable surface, discovered in 1858 by Johann Listing and Augustus Möbius. The most famous example is the so-called *Möbius strip*, which already appears in third century Roman mosaics, and is obtained by taking a rectangular paper band, holding one end (in the sense of the length), turning the other end 180 degrees, and finally gluing the two short ends together (without the turn one would obtain a cylinder). The Möbius strip has only one side and only one edge (fig. 2.10). Moreover, it is not orientable, in the sense that on it one cannot distinguish the clockwise and anticlockwise directions (or the right and left hands). Indeed, a top that spins in a certain direction while traveling once along the strip returns to its starting point spinning in the opposite direction.

The works of Riemann in 1857, Möbius in 1863, and Felix Klein in 1882 together led to the proof that every closed 2-dimensional surface is equivalent, from a topological point of view, to exactly one of the surfaces belonging to two infinite families. The first family consists of the sphere and the (orientable) surfaces that are obtained by adding to the sphere a finite number of (cylindrical) handles. A particularly interesting case is the sphere with only one handle, which is equivalent

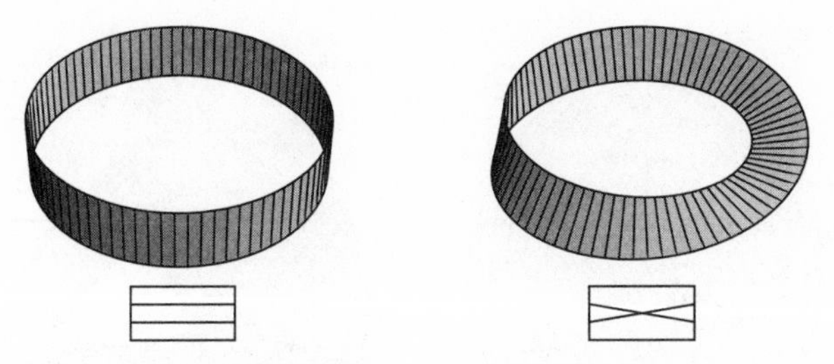

Fig. 2.10 Cylinder and Möbius strip.

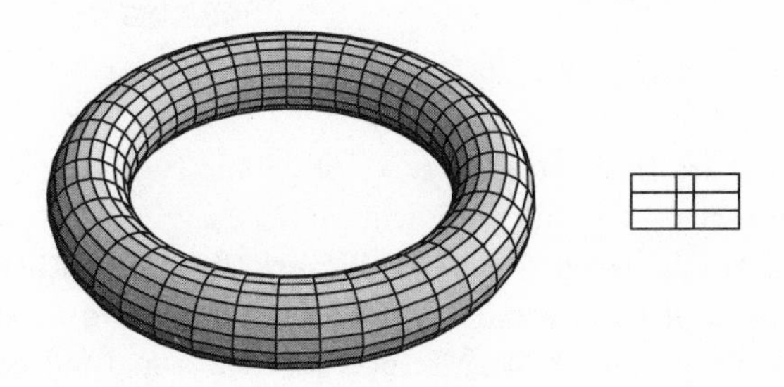

Fig. 2.11 Torus.

Fig. 2.12 Classification of the orientable 2-dimensional surfaces.

to the ring-shaped surface called *torus* (fig. 2.11). In particular, the 2-dimensional orientable surfaces are completely determined by the number of their holes (fig. 2.12). The second family consists of the (nonorientable) surfaces that are obtained by punching on the sphere a finite number of holes and replacing them with Möbius strips (this can be done because the strip has only one edge). Two particularly interesting cases are the spheres with one or two additional strips. They are equivalent, respectively, to the surfaces known as the *projective plane* and the *Klein bottle* (fig. 2.13).

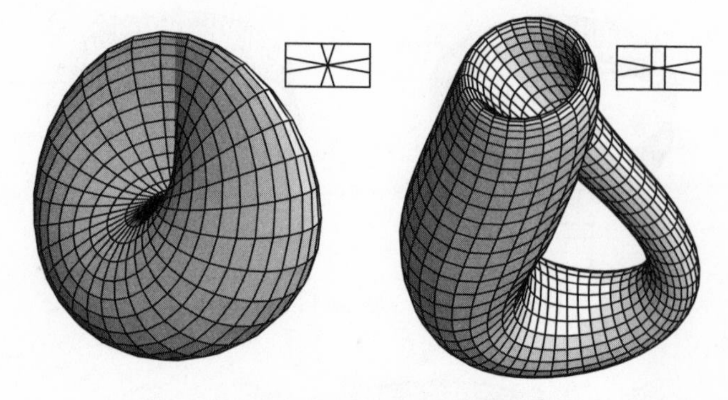

Fig. 2.13 Projective plane and Klein bottle.

There are three possible types of geometry for a 2-dimensional surface: the (usual) Euclidean, the hyperbolic, and the spherical (the latter differs significantly from the other two, because in it there are no parallel lines, since two maximum circles always meet). With respect to their associated geometries, the surfaces in the two families are divided as follows: to the sphere and the projective plane we can assign a spherical geometry; to the torus and the Klein bottle, a Euclidean geometry; and to all the remaining surfaces, a hyperbolic geometry.

After the classification of the surfaces in 2 dimensions was achieved, it was a natural step to attempt the classification of the 3-dimensional ones. This was begun in the 1970s by William Thurston and, even if the work is not yet completed, he was awarded the Fields Medal in 1983 for his efforts. He showed that in the 3-dimensional case there are not just three possible geometries but eight. These are those of the Euclidean space, of the hyperbolic space, of the hypersphere, of the hypercylinders with spherical section, of the hypercylinders with hyperbolic section, and three others (two of which correspond to defining in the Euclidean space a distance different from the usual one). To make matters worse, not all 3-dimensional surfaces admit only one of these geometries; it then becomes necessary, in general, to cut the surface up into pieces and assign different geometries to the various pieces. Fortunately, as Milnor showed in 1962, 3-dimensional surfaces may be decomposed in canonical pieces in an essentially unique way, through appropriate 2-dimensional cuts. Thus, there "only" remains to assign geometries to the canonical pieces, and this has already been done for many (although not yet for all) 3-dimensional surfaces. As in the case of the surfaces in 2 dimensions, the hyperbolic geometry continues here again to claim the lion's share.

We have already mentioned, in our discussion of the exotic manifolds, a topological classification of the surfaces in 4 dimensions due to Michael Freedman, who for this result was awarded the Fields Medal in 1986. For the surfaces in 5 or more dimensions, a classification can be obtained from homotopy theory, which we shall examine later on. The tridimensional case thus remains the only one to be completed, but this is not the end of the story.

There is in fact an important subclass of multidimensional surfaces, consisting of the *algebraic manifolds* (real or complex) defined by systems of algebraic equations. The complex *1-dimensional algebraic manifolds* (or *curves*) are certain particular real surfaces, and their topological classification then follows from the above general classification in terms of the number of holes.

A classification of the complex *2-dimensional* (or real 4-dimensional) *algebraic manifolds* (or *surfaces*), was one of the spectacular results of the Italian school of geometry of Guido Castelnuovo, Federigo Enriques, and Francesco Severi, obtained between 1891 and 1949. In certain cases, as for instance in that of the so-called *general type* surfaces, the Italians obtained a complete result. In other cases, as in that of the so-called *irregular surfaces*, the proofs remained incomplete because of a lack of the necessary technical tools, which were only developed in the fifties by Kunihiko Kodaira. For this work he received the Fields Medal in 1954 and the Wolf Prize in 1984–

85. The classification theorem for 2-dimensional algebraic manifolds is today known as the Enriques-Kodaira theorem.

The more difficult study of complex *3-dimensional* (or real 6-dimensional) *algebraic manifolds* was initially undertaken by Corrado Segre, but in this case the lack of suitable technical tools was an even bigger obstacle and did not allow the Italian school to advance beyond some remarkable intuitions and conjectures. The development of the necessary technology and the classification of 3-dimensional algebraic manifolds was instead one of the spectacular results obtained by the Japanese school of geometry of Heisuki Hironaka, Shing Tung Yau, and Shigefumi Mori. For this work they were awarded the Fields Medal in 1970, 1983, and 1990, respectively. In particular, Hironaka showed how to solve the singularities of a manifold by transforming the latter into another manifold without singularities. Yau characterized the *Calabi-Yau manifolds*, which not only are an important part of the classification but, as we shall later see, have also found some unexpected applications in string theory. Mori stated and brought to a conclusion the so-called *minimal model program*, on which the classification is based.

2.14. Number Theory: Wiles's Proof of Fermat's Last Theorem (1995)

In 1637 Fermat read Diophantus's *Arithmetic*, a monumental third-century treatise, and noted on the margin the following observation: "Dividing a cube into two cubes, or in general an *n*th power into two *n*th powers, is impossible if *n* is larger than 2. I have found a truly remarkable proof of this fact, but the margin is too narrow to contain it." For the cubes, this observation had been anticipated in 1070 by Omar Khayyâm, mathematician and poet, author of the *Robâ'iyyât*. In its general form, the claim is known as *Fermat's last theorem*, and for

350 years it has been one of the most famous problems in mathematics.

Fermat required that n be greater than 2, for already the Babylonians, and later the Pythagoreans, knew that there are squares that can be written as the sum of two squares, for example $3^2 + 4^2 = 5^2$, that is, $9 + 16 = 25$.

A proof of the theorem for $n = 4$ was discovered among Fermat's papers. The proof employs an ingenious method known as *infinite descent*. It consists in first assuming, by contradiction, that there is a solution, and then showing that there must be another solution in which the three numbers are not larger than those of the previous one, and with at least one of them strictly smaller; this leads to an impossible infinite regression.

In the course of time, the best mathematicians worked on the problem and confirmed the theorem for various particular cases: $n = 3$, Euler in 1753; $n = 5$, Dirichlet and Legendre in 1825; $n = 7$, Lamé in 1839; every n less than 100, Kummer between 1847 and 1857. By 1980, Fermat's assertion had been verified to be true for every n less than 125,000, but a general proof of the theorem was still missing.

The first truly general result was obtained in a rather indirect way. The starting point is the observation that Fermat's theorem asks for *integer* solutions of equations of the form

$$a^n + b^n = c^n.$$

Since it follows from this that

$$\left(\frac{a}{c}\right)^n + \left(\frac{b}{c}\right)^n = 1,$$

it is then a question of finding *rational* solutions of equations of the form

$$x^n + y^n = 1.$$

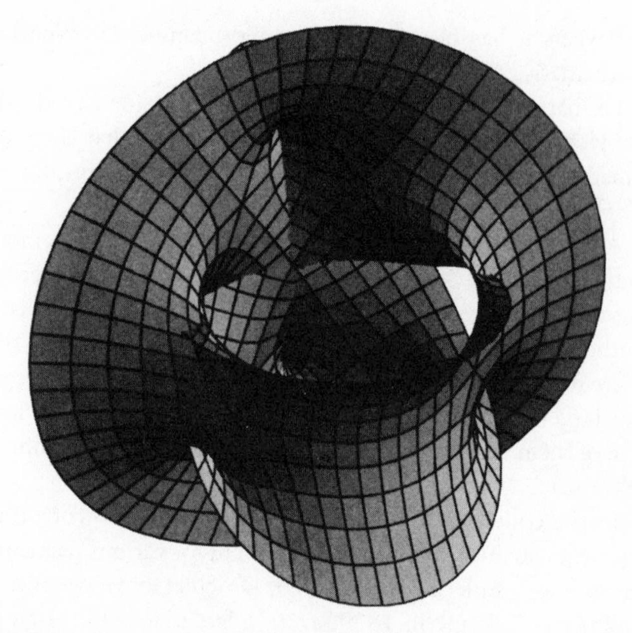

Fig. 2.14 The surface corresponding to the equation $x^3 + y^3 = 1$.

When x and y are considered to vary over the real numbers, these equations define a curve, and they define a surface when x and y are regarded as complex variables. These surfaces may then be classified according to the number of their holes. For example, for $n = 2$, there are no holes, because the curve defined by the equation is a circle and the surface is a sphere; and there are infinitely many rational solutions, which Diophantus already knew how to describe.

For n greater than 2 there are holes: one for $n = 3$, three for $n = 4$, six for $n = 5$, and so on (fig. 2.14). Of course, as the number of holes increases so does the complexity of the surface, and the possibility of finding simple (i.e., rational) solutions decreases.

Besides the one mentioned above, another type of equation had in the meantime turned out to be particularly interesting, the so-called *elliptic curves*, to which we have already referred. For these curves the number of holes of the corresponding surface is one, and also in this case it is possible to have an infinite number of rational solutions.

In 1922 Leo Mordell proposed the *Mordell conjecture*: the only type of equation that admits infinitely many rational solutions are those that define surfaces with either no holes or a single hole. This means that, if the Mordell conjecture is true, Fermat's theorem is *almost* true, because for all n greater than 3 (the case $n = 3$ having already been solved by Euler) the equation defines a surface with more than one hole, and therefore it can have, at most, a *finite* number of rational solutions.

In 1962 Igor Shafarevich came up with his own *Shafarevich conjecture*: under certain conditions, it is possible to find the solutions of an equation by first restricting the integers to values that do not exceed certain prime numbers, solving these (finite) particular cases of the equation, and finally putting these solutions together to obtain a solution of the original equation. In other words, the method seeks to recover the solutions from the knowledge of their remainders after division by various prime numbers.

A connection between the two conjectures was found in 1968 by Parshin, who showed that Mordell conjecture follows from that of Shafarevich. The latter was proved in 1983 by Gerd Faltings, who received the Fields Medal in 1986 for this result. His proof makes an essential use of Deligne's solution of another conjucture, due to Weil, that we shall examine later.

The proof of the Mordell conjecture is such an interesting result that it was hailed as "the theorem of the century," but it does not appear to be of much help in proving Fermat's theorem, for even a single rational solution of the equation

$$x^n + y^n = 1$$

would provide an integer solution of the equation

$$a^n + b^n = c^n,$$

and therefore an infinite number of these (obtained by multiplying the first solution by a constant). But in fact in 1985 Andrew Granville and Roger Heath-Brown succeeded in deriving from Faltings's theorem the validity of Fermat's theorem for infinitely many prime exponents—actually, for *almost all* exponents, from the perspective of measure theory.

The proof of Fermat's theorem for *all* exponents greater than 2 was obtained by following yet another indirect path, passing through the so-called *Taniyama conjecture.* This time, the starting point is the observation that the equation

$$x^2 + y^2 = 1$$

can be parametrized by the trigonometric functions sine and cosine, which satisfy the fundamental equation

$$(\sin \alpha)^2 + (\cos \alpha)^2 = 1.$$

Hence, solving Fermat's equation for $n = 2$ amounts to finding an angle whose sine and cosine are both rational numbers. In a similar way, the so-called hyperbolic sine and hyperbolic cosine parametrize the equation

$$x^2 - y^2 = 1.$$

Going from the quadratic equations that define the conics to the cubic equations, Taniyama conjectured in 1955 that certain *modular functions*, more general than the trigonometric functions, parametrize in a similar way an arbitrary elliptic curve.

The connection between the Taniyama conjecture and Fermat's theorem was noticed in 1985 by Gerhard Frey, who proposed to associate the elliptic curve

$$y^2 = x\,(x + a^n)\,(x - b^n)$$

to the Fermat equation

$$a^n + b^n = c^n.$$

Frey noticed that his elliptic curve had properties too good to be true. For instance, the discriminant that determines the existence of the roots of the polynomial

$$(x + a^n)(x - b^n) = x^2 + x(a^n - b^n) - a^n b^n,$$

namely

$$\Delta = \sqrt{(a^n - b^n)^2 + 4a^n b^n} = a^n + b^n = c^n,$$

is a perfect nth power. In 1986 Ken Ribet showed that Frey's curve cannot be parametrized by modular functions, which, differently put, means that Fermat's theorem follows from Taniyama's conjecture.

Now it was "just" a question of proving the conjecture. In 1995 Andrew Wiles managed to prove part of it, for a class of elliptic curves known as semistable, to which Frey's curve belongs. In so doing he solved one of the most famous open problems of modern mathematics. For this historical result Wiles obtained the Wolf Prize in 1995–96, but he could not receive a Fields Medal in 1998 because he had just turned forty.

In 1999 Brian Conrad, Richard Taylor, Christophe Breuil, and Fred Diamond completed Wiles's work by showing that the Taniyama conjecture is also true for the nonsemistable elliptic curves.

2.15. Discrete Geometry: Hales's Solution of Kepler's Problem (1998)

In 1600 Sir Walter Raleigh asked the mathematician Thomas Harriot for a formula to calculate how many cannon balls there are in a pile. The answer depends, of course, on how the balls are piled up, and Harriot asked himself what was the

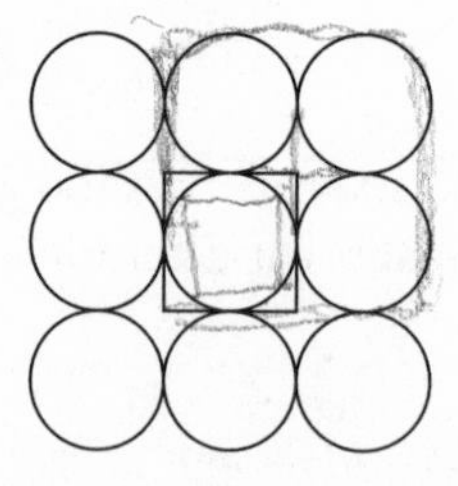
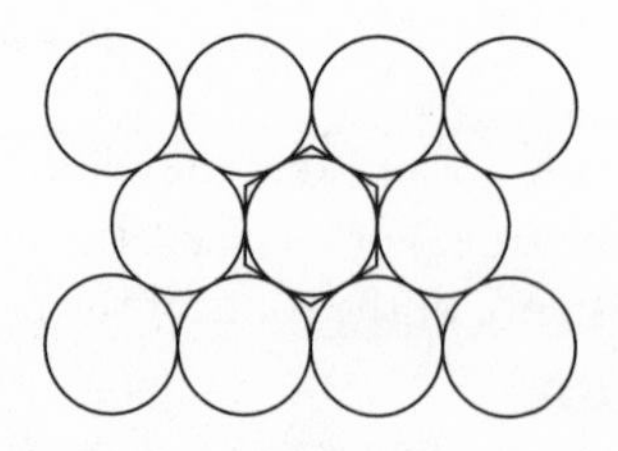

Fig. 2.15 Configurations of circles.

most efficient way to do this. In 1606 the problem was brought to the attention of the astronomer Johannes Kepler, who found that it was analogous to the problem of the formation of snow crystals, beehive cells, and pomegranate seeds. Kepler suspected that in all these cases the same mechanism was in action, by virtue of which spheres disposed in spatial grids of various shapes tend to completely fill the intermediate space as they grow.

In 1611 Kepler restated the underlying mathematical problem as follows: find the configuration of spheres of a given radius having maximum density—i.e., the minimum (in the limit) ratio between the total volume of the spheres and that of the space that contains them. A similar problem in the plane asks for the configuration of circles of a given radius having maximum density, this time with respect to the area.

The two obvious configurations to consider for the circles are the square and the hexagonal (fig. 2.15), and Kepler found that their densities are, approximately, 0.785 and 0.907, respectively. The hexagonal configuration is therefore the best of the two, as one can also see by simple observation. But this does not solve the problem, which asks for the best possible configuration!

In 1831 Gauss showed that the hexagonal is the best one among all the lattice configurations, that is, such that the centers of the circles form a *planar lattice*, i.e., a symmetric configuration of parallelograms. In 1892 Axel Thue claimed to

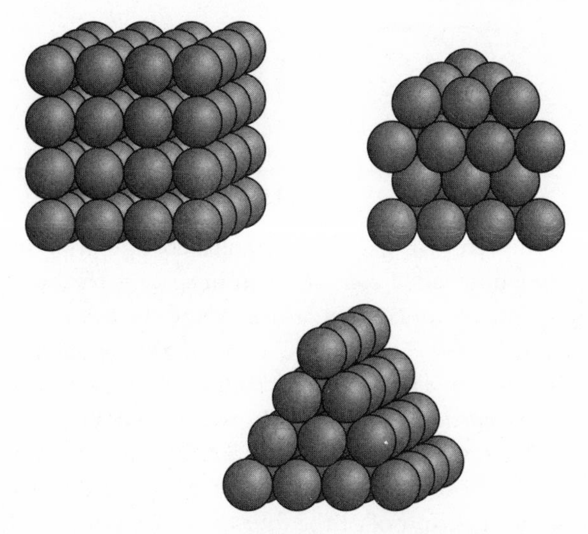

Fig. 2.16 Configurations of spheres.

have proved that the hexagonal configuration is the absolute best, but his proof was not published until 1910.

In space, the four obvious configurations to consider are those obtained by placing layers of spheres on top of each other. There are two choices for the configuration of the horizontal layers (square or hexagonal), and two choices for the vertical layers (with the centers of the spheres aligned or not). In fact, the above four configurations are only three, for when horizontal layers are disposed with the centers not aligned, both square and hexagonal layers produce the same configuration (fig. 2.16).

Kepler calculated the densities of the square aligned, hexagonal aligned, and (square or hexagonal) nonaligned configurations. They are approximately 0.524, 0.605, and 0.740, respectively. The nonaligned configuration is therefore the best of the three. And, in fact, it is the one that is spontaneously used at the store when arranging fruit for display. But, once again, this is not a solution to the mathematical problem.

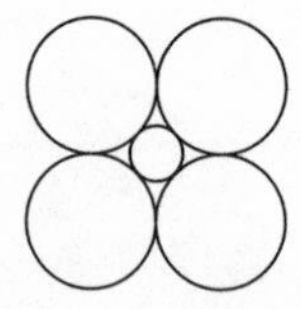

Fig. 2.17

Gauss showed that, like the hexagonal configuration in the plane, the non-aligned configuration in space is the best one among all the lattice configurations, where the centers of the spheres form a spatial lattice, i.e., a symmetrical configuration of parallelepipeds. The general case constituted the third part of *Hilbert's eighteenth problem.* It was solved in 1998 by Thomas Hales, who proved that the nonaligned configuration is indeed the most efficient one. The structure of the proof is reminiscent of the proof of the four-color theorem, to which we shall return later. In both cases, it is a question of reducing the number of configurations to be checked until it is sufficiently small for a computer to do the job. In Hales's proof, the reduction required 250 pages, and the computer program 3 gigabytes.

As the number of dimensions increases, so does the interest of the problem. In 2 dimensions, we can place four circles of radius 1 inside a square of side 4, and there is room left at the center for a small circle of radius $\sqrt{2} - 1 \approx 0.41$. In 3 dimensions, we can place eight spheres of radius 1 inside a cube of side 4, and there is room left at the center for a small sphere of radius $\sqrt{3} - 1 \approx 0.73$ (fig. 2.17). More generally, in n dimensions we can place 2^n hyperspheres of radius 1 inside a hypercube of side 4, and there is room left at the center for a small hypersphere of radius $\sqrt{n} - 1$.

The radii of the small hyperspheres that can be inserted between the hyperspheres keep increasing with the number of dimensions, as can already be seen when the dimension increases from 2 to 3. In 9 dimensions, the small hypersphere has radius $\sqrt{9} - 1 = 2$ and therefore it touches the faces of the hypercube, and for *n* greater than 9 it actually bursts through them!

The problem of the best lattice configuration of multidimensional hyperspheres has been solved up to dimension 8. But the lattice configuration is not always the one with the best density. This happens, for instance, in 10 dimensions, as John Leech and N.J.A. Sloane showed in 1971.

The case of dimension 24 is particularly interesting. In 1965 Leech constructed a configuration, known as the *Leech lattice*, which is probably the best among the lattice configurations, and in which each hypersphere touches 196,560 others (in the nonaligned configuration in 3-dimensional space, each sphere touches twelve others). From the study of this lattice, John Conway derived, in 1968, three of the twenty-six sporadic groups used in the classification theorem for finite simple groups.

The problem of finding the configuration of maximum density for hyperspheres in multidimensional spaces is today a question of considerable importance in the transmission of messages, in particular for data compression and error correction. Binary strings of *n* symbols may be seen as the vertices of a hypercube in *n* dimensions. To avoid transmission errors, one would like to prevent that vertices which encode messages be adjacent to each other. A configuration of hyperspheres with maximum density maximizes the number of encoded messages while minimizing the possibility of errors. The Leech lattice was discovered precisely in the context of this type of problem.

CHAPTER 3 APPLIED MATHEMATICS

Mathematics, like the Roman god Janus, has two faces. One is turned inward, toward the human world of ideas and abstractions, while the other looks outward, at the physical world of objects and material things. The first face represents the purity of mathematics, where the attention is unselfishly focused on the discipline's creations, seeking to know and understand them for what they are. The second face constitutes the applied side of mathematics, where the motives are interested, and the aim is to use those same creations for what they can do.

The applications of mathematics have been a permanent characteristic of its history, from Egyptian and Babylonian times to the industrial revolution, and every branch of classical mathematics was, at the outset, stimulated by practical problems: arithmetic by accounting problems; geometry by agricultural problems; and mathematical analysis by problems in physics. With the passage of time, these areas were constantly spurred by pragmatic and utilitarian motivations, which have also contributed to their theoretical development, often with unexpected fallouts.

Twentieth-century mathematics is no exception in this respect, and many of its new branches were born precisely as a consequence of external solicitation, for the purpose of solving problems arising from the real world. Some of these motivations come from well-established scientific domains such as physics, which inspired, if not the birth, certainly the growth of tensor calculus, functional analysis, and knot theory, all essential tools in general relativity, quantum mechanics, and string theory.

Other motivations originated, instead, from domains that became scientific not until the twentieth century, precisely when the discovery of suitable mathematical tools allowed experts to treat and solve some of their fundamental problems. Typical examples are economics and biology. Game theory, general equilibrium theory, and optimization were created to solve problems in the first field, while problems in biology, long considered inaccessible, can today be tackled using knot theory.

The mathematical tools we have just mentioned, and which we shall discuss in more detail later on, are pinnacles of technical sophistication. But this technical complexity is not essential for a mathematical argument to have a dazzling effect, provided its absence is compensated by a proper dose of philosophical sophistication. Before proceeding further we shall show, with three examples from the above fields, how even the most elementary mathematics may suffice, if cleverly used, for solving some important foundational problems in science.

The first of these problems concerns the notion of *reality* in physics, which was called into question by the discovery of quantum mechanics, and more precisely by the description of subatomic phenomena in terms of wave functions. Due to difficulties in its interpretation, Niels Bohr proposed to consider the theory as describing not hypothetical physical particles but only the outcome of experiments carried out on measuring devices. According to Bohr, the notion of reality that had been developed in the past to describe the macroscopic world ceases to be meaningful at the microscopic level.

Such an idealistic interpretation of the new physics of course encountered some strong resistance, in particular from Albert Einstein. He continued to believe for the rest of his life that a realistic description of subatomic phenomena could be found, and of which quantum mechanics would have been only an approximation. In 1935 he proposed a famous thought experiment—known as the Einstein, Podolski, Rosen experiment, after its three authors—which demonstrated the incompleteness of quantum mechanics.

In 1964 John Bell found a version of the thought experiment that could be empirically tested, and whose outcome was unexpected. Consider a ray of light that goes through two polarized filters. Quantum mechanics predicts, and experience confirms it, that after the ray of light has gone through the first filter, the fraction of its photons that also crosses the second filter is $\cos^2(\alpha)$, where α is the angle between the directions of polarization of the two filters.

We now consider what happens when each of the two filters is placed either vertically, at 60 degrees, or at 120 degrees. If both filters have the same direction—which occurs in one-third of the nine possible cases—the second filter lets through all photons of the ray coming from the first filter. But if the filters have different directions—which occurs in the remaining two-thirds of the cases—they always form a 60-degree angle, and the second filter lets through $(1/2)^2 = 1/4$ of the photons coming out of the first one. Hence, on average, only $1/3 + 2/3 \times 1/4 = 1/2$ of the photons go through.

What Bell discovered is that these experimental results are in contradiction with the hypothesis that the photons can realistically be thought of as particles that arrive at the filters already polarized in a particular direction. If this were the case, when the filters have the same direction, the same photons would pass through both filters. If instead each filter is polarized in any one of the three directions, at least five-ninths of the photons coming from the first filter should pass through the second one, which is more than one-half of them. In effect, in the three cases in which the filters have the same direction, the same photons pass through both of them; and if a photon goes through the filters placed in two different directions, it should also go through when these two directions are interchanged, that is, in two other cases.

A simple arithmetical calculation has thus shown that the hypothesis of naive realism is contradicted by experimental results. More sophisticated versions of Bell's theorem, confirmed by the famous experiments carried out by Alain Aspect in 1982, show that while it is possible to interpret quantum mechanics realistically, this cannot be done by keeping intact our usual notion of reality at the macroscopic level. In particular, we can no longer assume that objects separated in space cannot interact instantaneously, and the existence of holistic connections, which are not part of the Western cultural tradition, must therefore be postulated.

The second foundational problem is about the notion of *social choice* among several alternatives, based on the knowledge of individual preferences. The problem arises in the most diverse situations, from the choice of candidates in a political election, to that of an economic plan by a board of directors.

One of the difficulties with this problem was discovered in 1785 by Marie Jean Antoine Nicolas de Caritat, better known as the Marquis of Condorcet, and may be illustrated with a practical example. In the 1976 American election, Jimmy Carter defeated Gerald Ford, who had won the Republican nomination over Ronald Reagan. But according to the polls, Reagan would have beaten Carter—as he actually did in 1980, although under different political circumstances. Thus, a paradoxical situation foreseen by Condorcet had taken place: in an electoral system in which candidates are chosen through successive votes, one candidate against another, the winner may depend on the order in which the voting rounds take place. For example, Ford could have won, if a vote to choose between Carter and Reagan would have been held first, followed by another vote between the winner (Reagan) and Ford.

The obvious question is whether the electoral system can be reformed in order to avoid a situation similar to the above. The surprising negative answer was found by Kenneth Arrow in 1951 and was the origin of the theory of social choice for which Arrow received the Nobel Prize in economics in 1972.

Arrow's theorem establishes that no electoral system can satisfy all of the following principles: individual freedom (i.e., every elector can vote for the candidate of his or her choice); decision by vote (the result of the election depends only on the votes received by each candidate); unanimity (a candidate receiving all the votes wins); and rejection of dictatorship (no single elector can by himself or herself decide the outcome of every election).

Clearly, the hypotheses underlying Arrow's theorem are usually considered mandatory in any democratic system, and for this reason it is said, metaphorically, that Arrow has shown that democracy does not exist. What is interesting from our perspective is that the proof of the theorem is a mathematical argument, based upon a simple axiomatization of the conditions behind Condorcet's paradox. This shows that mathematics can also be applied to the humanities, a domain that might have appeared at first sight to be unsuitable to formal analysis.

Our last foundational problem is about the notion of *self-reproduction*, characteristic of living organisms. In 1951, as John von Neumann was working on the theory of cellular automata, he considered the problem of building a machine capable of reproducing itself. He solved the problem mathematically, inspired by a technique used in the theory of computability.

Let us consider a machine B that behaves as a universal builder, in the sense that it is capable of building any machine M of a certain type from a description m of the given machine. In particular, B can build its own copy from a description b of itself, but this does not really constitute self-reproduction. In effect, starting with the system formed by the machine *B* and its description *b*, a copy of *B* is obtained, but a copy of its description *b* is missing.

To circumvent this difficulty let us consider a machine *P* that behaves as a universal photocopier, capable of reproducing any given description *m*. By combining machines *B* and *P* we obtain a new machine *A*, which, from a description *m*, makes a copy of it, constructs *M*, and produces as output both *M* and a description *m* of it. The machine *A* together with its own description *a* is now a self-reproducing system, because it constructs the machine *A* and a description *a* of it.

While the above process was thought of in the context of mechanical reproduction, in 1953 Francis Crick and James Watson found that von Neumann's argument also provides a molecular model of biological reproduction, in a work that was rewarded with the 1962 Nobel Prize in medicine. More precisely, the description *m* plays the role of a gene, or DNA segment, which encodes the necessary information for reproduction. *P*, a special enzyme known as RNA polimerase, can duplicate the genetic material in an RNA segment. *B*, a set of ribosomes, builds proteins according to the information provided by this segment. *A* is a self-reproducing cell.

Of course, not only is the model oversimplified, but it also completely bypasses the chemical "details" of the mechanism, in particular the famous double-helix structure of DNA discovered by Crick and Watson. The interesting thing, from our point of view, was to show how the general plan for reproduction could be discovered theoretically, and was actually discovered in practice, through a simple use of logical techniques.

After this introduction that illustrated some applications of elementary mathematics to foundational problems, we shall now present some applications of higher mathematics to problems of a more specifically scientific nature.

3.1. Crystallography: Bieberbach's Symmetry Groups (1910)

The commandment that in the Christian tradition has been shortened to "you shall have no other Gods besides me," in its original formulation (*Exodus*, 20:3–6; Deuteronomy, 5:7–10) continued: "You shall not carve idols for yourselves in the shape of anything in the sky above or on the earth below or in the waters beneath the earth."

The interdiction of a figurative art was taken seriously by Arabs and Hebrews, who developed a purely abstract and geometric art and explored the possible types of mural decorations. The best result in this field was achieved in the fourteenth century, with the tiling of the Alhambra in Granada (fig. 3.1).

The number of possible mural decorations is of course unlimited, but there is only a limited number of different types. From a mathematical point of view, the symmetries that such decorations exhibit may be classified on the basis of the possible combinations (more precisely, of the possible symmetry groups) of transformations that leave them invariant: translations along a line, reflections with respect to a line, and rotations about a point.

In 1891 E. S. Fedorov showed that there are only seven different types of symmetry groups for linear friezes, such as ornamental moldings and plinths (fig. 3.2), and 17 for planar decorations, such as floors and tapestries (fig. 3.3). In addition, the planar groups can only have rotational symmetries of 180, 120, 90, and 60 degrees, that is, they may be either axial, triangular, square, or hexagonal. Almost all of these types have actually been used in the decorations of the Alhambra, as well as in those of various other civilizations, from the Egyptians to the Japanese.

If the most common symmetric objects in two dimensions are mural decorations, the best known in space are crystals. Crystallography was precisely one of the first domains of appli-

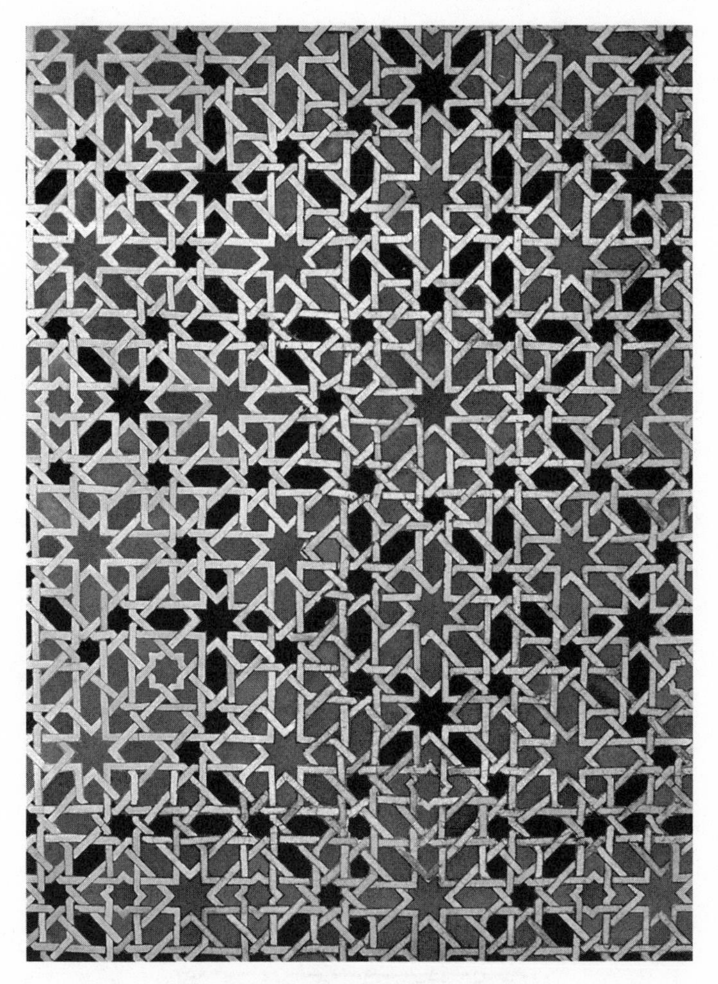

Fig. 3.1 A tiling in the Alhambra.

cation of group theory, beginning in 1849 with Auguste Bravais. In 1890, before showing the analogous result for the types of symmetry groups in the plane, Fedorov had already proved that there exist only 230 different types of spatial symmetry groups.

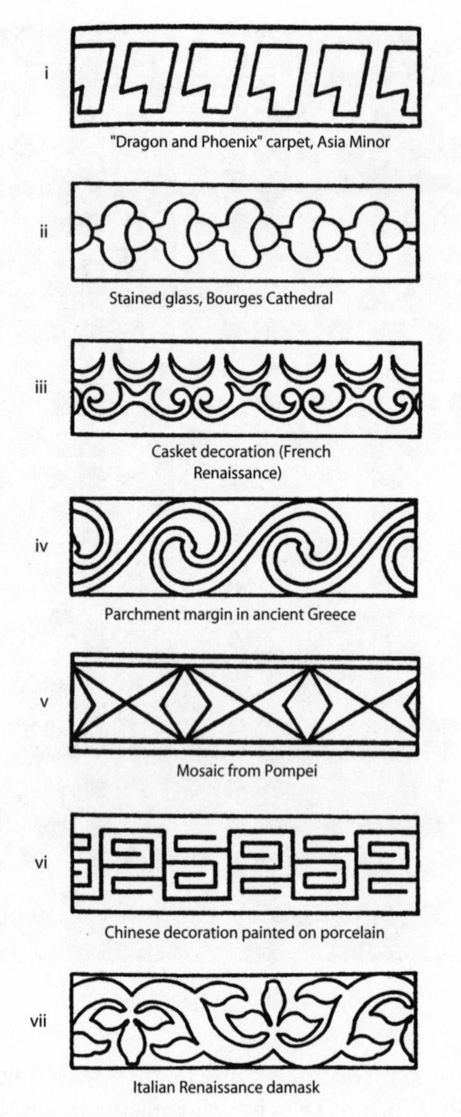

Fig. 3.2 The seven linear symmetry groups.

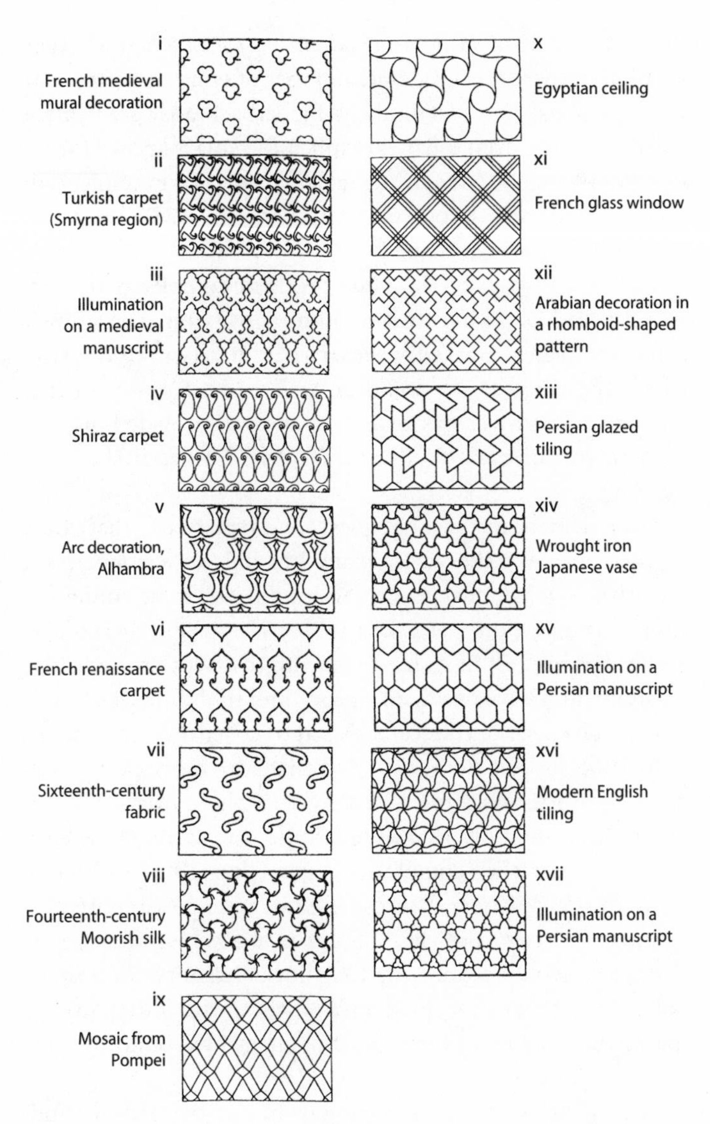

Fig. 3.3 The seventeen planar symmetry groups.

The first part of *Hilbert's eighteenth problem* asked whether, for each *n*, there is only a finite number of symmetry groups in *n* dimensions. In 1910 Ludwig Bieberbach answered in the affirmative, but even today no explicit formula is known for the number of such groups in the general case. For example, only in the 1970s was it proved that there are 4783 symmetry groups in four dimensions.

The second part of the same problem completed the first one. Instead of asking for the number of symmetric ways of tiling the plane, it asked whether there is a type of tile with which the whole plane could be covered in a nonsymmetric manner. The answer was again yes, and was provided in 1935 by Heesch (an example due to Maurits Escher is shown in fig. 3.4).

More demanding is the request for a type of tile that could be used to cover the whole plane but only in a nonperiodic way, that is, without repeating to infinity the same configuration. The question was asked in 1961 by the logician Hao Wang. His interest in it stemmed from the fact that a negative answer would have provided a procedure to decide whether or not any given set of tiles can be used to cover the entire plane.

In 1966 Robert Berger showed that such a decision procedure does not exist, and therefore that there exist tiles that cover the plane only in a nonperiodic way. Berger's original example was rather complicated, involving 20,246 different tiles. In 1974 Roger Penrose found a simple example requiring only two tiles (fig. 3.5). It is not known if there are examples involving only one tile (but an example of a single polyhedron that can be used to cover the entire 3-dimensional space, and only in a nonperiodic manner, was found in 1993 by John Conway).

Penrose's example is interesting from a mathematical standpoint because it exhibits a pentagonal rotational symmetry (fig. 3.6) that no symmetric planar tiling can possess. The result also became relevant in physics when, in 1984, the crystallogra-

Fig. 3.4 Maurits Escher, *Ghosts*, 1971.

pher Daniel Schechtman discovered an aluminum and manganese alloy having a molecular structure whose surface exhibits a symmetry of the same type, which no crystalline structure can have. These type of structures have been called *quasicrystals*.

The discovery of quasicrystals demonstrates that group theory is not the ultimate tool for the description of nature, and

Fig. 3.5 Penrose tiles.

Fig. 3.6 Penrose symmetry.

that some more general theory is therefore necessary. For this reason the study of the properties of quasicrystals and the search for a classification of their structures—in particular the *quasicrystallographic groups*—has attracted mathematicians like Sergei Novikov and Enrico Bombieri, Fields medalists in 1970 and 1974, respectively.

3.2. Tensor Calculus: Einstein's General Theory of Relativity (1915)

The fact that the Earth has long been believed to be flat shows intuitively that the larger the radius of a sphere, the smaller its curvature. Formally, the curvature of a circle is defined as the reciprocal of its radius. For more complex curves, curvature was defined by Newton in 1671 by considering at each point of the curve the curvature of the circle (called the osculating circle—from the Latin *osculum*, kiss) that approximates the curve at that point.

The curvature of a surface was defined by Gauss in 1827, by considering, at each point, the product between the minimum and the maximum curvature of the curves (sections) obtained as the intersection of the surface with the planes perpendicular to the tangent plane and passing through that point. For example, the sphere has the same curvature as that of its great circles, which are precisely its sections; and the cylinder has zero curvature, because one of its sections is a straight line.

In order to calculate the curvature of a surface as defined by Gauss, it is necessary to be able to perform measurements from the outside, that is, from the space containing the surface. But Gauss discovered that it is also possible to calculate the curvature by means of measurements performed only on the surface itself and, in particular, to conclude that the earth is round without having to observe it from outer space.

Gauss then proved a theorem so satisfying that even he, famous for being extremely demanding, called it *theorema egregium*, or "excellent theorem": surfaces that possess an intrinsic geometry—i.e., such that figures drawn on the surface can be moved around without being deformed—are precisely those with constant curvature. On such surfaces, the counterparts of the straight lines are the so-called *geodesics*, or lines of minimal distance between two points. On a sphere, for example, the geodesics are the arcs of great circles; and on a cylinder, they are the curves obtained by joining the two points with a line segment, after having cut out the cylinder in the direction of its axis and spread it flat on a plane.

In the plane, the only curves with constant curvature are the straight line, of zero curvature, and the circle, which has constant positive curvature. But Gauss discovered that there are also surfaces with constant negative curvature. The *pseudo-*

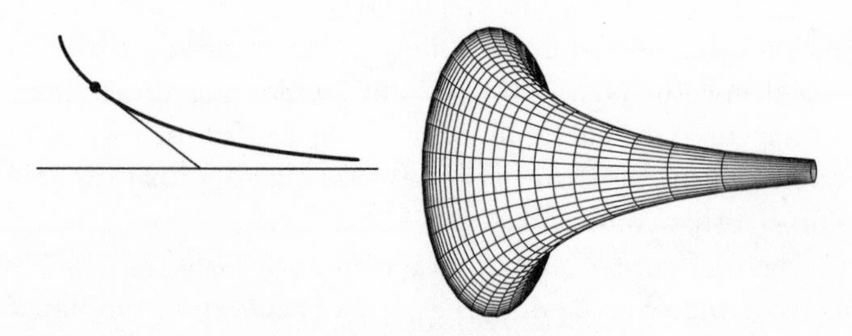

Fig. 3.7 Tractrix and pseudosphere.

sphere, for instance, generated by rotating around its asymptote a curve known as a *tractrix* (fig. 3.7), may be described as the path of a stone being dragged by pulling a string of fixed length while one walks along a straight line (i.e., the distance between the point of contact of a tangent to the curve and the intersection of this tangent with the asymptote is constant).

In 1854 Riemann extended the notion of curvature to his manifolds, which cannot always be embedded in Euclidean space, and he also determined the geometry of the manifolds with constant curvature. This geometry is Euclidean if the curvature is zero, spherical if the curvature is positive, and hyperbolic if the curvature is negative. In particular, the pseudosphere represents a model of a portion of the hyperbolic plane in Euclidean space (only a portion, because the pseudosphere has a hole but the hyperbolic plane does not). It was precisely while working on this partial model that Beltrami obtained the first complete model of the hyperbolic plane already mentioned in chapter 2.

Besides their role as models of mathematical geometries, Riemann manifolds may also be considered models of the physical world. The first to propose such a possibility was Gauss, who carried out geographical measurements to find out whether the geometry of the universe was really Euclidean, as had always been thought, or not.

The only quantities that are geometrically relevant are those, such as distance, that are independent of any given coordinate system. The same is true for the laws of physics. Since these are generally expressed in differential form, in order to apply Riemannian geometry to physics it was necessary to undertake first a study of the invariance of differential equations with respect to coordinate changes on Riemann manifolds.

The tool developed with this goal in mind by Gregorio Ricci Curbastro, beginning in 1892, was called *tensor calculus*. Tensors are quantities that are transformed in such a way that their components, in a new coordinate system, are linear combinations of their components in the old system, with coefficients given by the derivatives of the transformation. Ricci also defined algebraic operations (sum and product) as well as differential operations (covariant differentiation) on tensors, thus making it possible to extend to Riemann manifolds the whole analytical apparatus already developed in the Euclidean case.

In 1901 Ricci and Tullio Levi-Civita expressed in tensor form, and hence in a form that is invariant with respect to a change of coordinates, several laws of physics. But the most interesting application is due to Albert Einstein, who in 1915 found in tensor calculus the ideal tool to describe his general theory of relativity.

The Riemann manifolds used by Einstein have four dimensions, three of which are spatial and one is temporal. For this reason they are referred to as models of *space-time*. The specific form of the manifold, and in particular its curvature, is determined by the distribution of matter in the universe, and free bodies move on the manifold along geodesics, like masses that roll down a slope along lines of minimal resistance.

Once gravitation has been reduced to geometry, a natural step is to search for a similar reduction of the other forces in physics. The first formulation of a theory that also included electromagnetism was found by Hilbert in 1915. He elegantly derived both Einstein's and Maxwell's equations from a single variational principle, according to the requirements of his *sixth problem*, which asked for an axiomatization of physics.

A different attempt was made by Hermann Weyl, who in 1918 described both gravitation and electromagnetism using an affine (non-Riemannian) four-dimensional manifold, instead of a metric (Riemannian) one. In such manifolds, while parallelism is independent of the coordinate system, this is not necessarily the case for distance. Hence, a new definition of a geodesics is required, for it can no longer be defined as a line of minimum distance. This request was already contained in *Hilbert's fourth problem*, which asked for a general study of the notion of geodesics. The solution proposed by Levi-Civita in 1917 was to define the geodesics as those curves whose tangents are all parallel to one another.

Even if Weyl's (as well as Hilbert's) theory was not satisfactory from the point of view of physics, it opened up the study of non-Riemannian manifolds in geometry. A satisfactory common treatment of both the gravitational and the electromagnetic fields still remains an open problem, and it belongs to the more general problem that seeks to unify all the forces into a *theory of everything*.

3.3. Game Theory: Von Neumann's Minimax Theorem (1928)

Life continually forces us to make choices at every level (personal, familial, social) and in every domain (moral, economic, political), in cases where we have an imperfect knowledge of the situation, of the behavior of others, and of the consequences of the various alternatives. The purpose of *game theory* is to develop a mathematical model of this decision process in a typically scientific manner, that is, by abstracting from real situations some elements that could be treated in a formal way.

A first significant example of such an analysis going back to 1651 can be found in Thomas Hobbes's *The Leviathan*. In his great work, the English philosopher advances the idea that human societies are alliances imposed by the need to stifle the violent character of human nature, based, on the one hand, on aggression against all and, on the other hand, on the fear of others; in other words, on our preference for not cooperating with others but for all others to cooperate with us. Through the social contract, individuals give up their right to exercise violence in exchange for the safety of being protected, and the resulting social order favors not only those who impose it, but everyone. The consequence of this social contract is therefore a change in the rules of the game.

A second significant example is found in a passage of the 1755 *Discours sur l'origine de l'inegalité parmi les hommes* by Jean-Jacques Rousseau. This time human societies are seen as the evolution of the temporary alliances people make to hunt large animals that could not be caught by an individual hunting alone. But while two people are, say, hunting a deer, one of them may happen to see a hare, which he could hunt all by himself. Thus appears the temptation to go after the hare, on the grounds that, while a deer is better than a hare, a hare is better than nothing. The temptation is made all the stronger by the thought that perhaps the other individual has also spotted a hare and given up hunting the deer.

Other examples come from real games, which is the reason for the name of the theory. They may be played not just for pleasure, like cards or chess, but also for training, as in the case of the *Kriegspiel.* This game, which used military maps and toy soldiers, was considered the inspiration for the winning strategies in the Prussian wars against Austria in 1866 and France in 1870, and in the Japanese war against Russia in 1905.

The first mathematical work in game theory was the article presented by Ernst Zermelo to the 1912 International Congress of Mathematicians. In his article Zermelo proved that chess (and, more generally, any game that cannot go on forever) is determined, in the following sense: either there is a strategy for the white to always win, or there is a strategy for the black to always win, or there is a strategy for each player that always results in a tie. This existence result is, however, nonconstructive, in that it does not establish which of the three alternatives actually occurs, and for this reason it has no practical application.

The foundations of game theory were laid down in 1921 by Emile Borel, who was also France's minister of the navy. He used the game of poker as an example and, among others, tackled the difficult problem of an analysis of bluffing. In addition, Borel posed the problem of determining in which cases an optimal strategy exists, and how to find it.

An application of Brouwer's fixed-point theorem allowed John von Neumann to prove, in 1928, the first deep theorem of the new theory. It establishes that in certain games known as *zero-sum* (i.e., in which one player's gain equals the other player's loss) and *perfect information* games (i.e., in which each player knows exactly the possible moves of the other player as well as their consequences) there exists a strategy that allows both players to minimize their maximum losses—hence the name *minimax.*

For every possible move, each player considers all the possible moves of the other player and the maximum loss that he or she could suffer as a consequence, and then plays the move resulting in the minimum of these losses.

This strategy, which minimizes the maximum loss, is optimal for both players if their respective minimaxes are equal in absolute value and of opposite signs. If the common minimax value is zero, then it is useless to play.

The minimax theorem was improved and extended several times by von Neumann to imperfect information games or to games with more than two players, for example. This latter case is more difficult due to the possibility of cooperation among some players, in the form of alliances or coalitions. Von Neumann's work culminated with the publication in 1944 of the classical text *Theory of Games and Economic Behavior*, written jointly with the economist Oscar Morgenstern.

The most satisfactory formalization of the notion of an optimal strategy is the concept of the *Nash equilibrium*, proposed in 1950 by John Nash. In the particular case of zero-sum games, this concept reduces to the von Neumann minimax. Nash showed that each noncooperative game with two or more players, not necessarily zero sum, admits an equilibrium, and for this work he obtained the Nobel Prize in economics in 1994.

In the two-player case, a Nash equilibrium is a situation in which neither player has regrets, in the sense that had each known in advance the behavior of the other player, they would have both played the way they did. In other words, the situation cannot be improved through unilateral individual acts, although in general it may be improved by means of some joint action.

It is pretty obvious that if a state is not an equilibrium, then it is not rational, for at least one player would have reason to believe that he or she could have done better. Being a Nash equilibrium is therefore a necessary condition for rational behavior, but it is not a sufficient one, because there are games in which the states of the Nash equilibrium are not at all rational.

A typical example is the *prisoner's dilemma*, proposed by Albert Tucker in 1950. Two suspects of a crime have been arrested and are being questioned in separate cells. If only one of them denounces the other, he will be rewarded by regaining his freedom, while his accomplice will be sentenced to 10 years in prison. If each of them denounces the other, both will go to prison for 5 years. But if neither denounces the other, both will be set free. The only Nash equilibrium in this case is for each to denounce his partner, but it is not rational, for it is certainly in their common interest not to speak up.

In the second half of the century game theory has played a central role in the analysis and resolution of conflicts, and it is commonly applied by military, economic, and political advisers to government officials in various industrialized countries, foremost in the United States.

3.4. Functional Analysis: Von Neumann's Axiomatization of Quantum Mechanics (1932)

Problems in mathematical physics lead naturally to differential or integral equations, in which an unknown function appears under the derivative or the integral sign. Methods for the solution of differential equations, first in ordinary derivatives and later in partial derivatives, were developed beginning in the late seventeenth century. The first explicit steps toward the solution of integral equations, a more challeging problem, were only taken in the early nineteenth century. The general theory of integral equations was initiated in the last decade of the nineteenth century by Vito Volterra, and fully developed in the first decade of the twentieth century by David Hilbert.

This development of mathematical analysis revealed the essential fact that in mathematics one often works not only with functions that operate on numbers, but also with *functionals* that operate on functions. For example, the operations of squaring or of taking the square root explicitly associate with a given number another number—its square or its square root. Similarly, the operations of differentiation and of (indefinite) integration associate with a given function another function—its derivative or its (indefinite) integral. Also, an equation implicitly defines one or more numbers—i.e., its solutions—and in an analogous way a differential or integral equation implicitly defines one or more functions—its solutions.

The difficulties in the treatment of these functionals, especially in the calculus of variations and in the theory of integral equations, led to the call for an abstract and independent theory that would bring their properties to light. Its name, *func-*

tional analysis, indicates that the theory is different from (real or complex) analysis, which deals instead with functions that operate on (real or complex) numbers.

The natural context for the development of real (or complex) analysis are the Euclidean spaces, whose points are identified with their Cartesian coordinates. In the case of an n-dimensional space, a point is identified with n numbers $x_1, \ldots, x_n$, and the distance from such a point to the origin is calculated using the Pythagorean theorem, with the expression

$$\sqrt{x_1^2 + \ldots + x_n^2}$$

In his study of integral equations, Hilbert had to work with functions that can be expressed as an infinite sum (called Fourier series), involving an infinite number of coefficients $x_1, x_2, \ldots$, and he discovered that for these functions to be part of his theory, the condition was that the sum

$$x_1^2 + x_2^2 + \ldots$$

be finite. But if this sum is finite, so is its square root. These sequences of numbers may therefore be thought of as the coordinates of points in a Euclidean space with an "infinite number of dimensions" in which the Pythagorean theorem still holds. In 1907 Erhard Schmidt and Maurice Fréchet introduced the *Hilbert space H*, whose elements are the points with infinitely many coordinates that satisfy the above condition.

Since the sequences were only one way for Hilbert to deal with functions, Schmidt and Fréchet also defined a functional space L^2, whose points are the functions (defined on some interval) satisfying a condition analogous to that imposed by Hilbert, namely, that the Lebesgue integral of their square be finite; hence its name L^2. The fact that the Hilbert space H and the functional space L^2 are really the same thing is the essence of the so-called representation theorem due to Friedrich Riesz and Ernst Fischer.

The above two spaces H and L^2 are both special cases of a large class of *Banach spaces*, introduced in 1922 by Stefan Banach. They proved to be the correct axiomatization of the properties needed to develop the theory of integral equations. In particular, the construction of solutions of these equations through successive substitutions, according to a technique already anticipated in 1832 by Joseph Liouville, turned out to be special cases of a general fixed point theorem due to Banach.

But the real triumph of functional analysis was not so much its success in the domain of integral equations but rather its unexpected and immediate application to quantum mechanics. The latter theory had originally been formulated, by purely heuristic motivations, using two completely different formalisms, which nevertheless turned out to be equivalent. One was Werner Heisenberg's approach using infinite matrices of observables in 1925, an achievement for which he received the Nobel Prize in 1932; and the other in terms of wave functions due to Erwin Schrödinger, who was awarded the Nobel Prize for this work in 1933.

During the winter of 1926, in the spirit of his *sixth problem*, Hilbert himself had been trying to extract from the two formalisms an axiomatic formulation that would be theoretically satisfactory and from which both approaches could be derived. His ideas did not work at that time because the theory of distributions that would have justified them had not yet been developed. But in 1927 his assistant John von Neumann reformulated Hilbert's ideas in terms of the spaces H or L^2. In the first case one obtains Heisenberg's version of quantum mechanics, and in the second Schrödinger's. Moreover, the equivalence between the two follows from the Riesz-Fischer representation theorem.

In von Neumann's final model, which culminated in 1932 in the classic *Mathematical Foundations of Quantum Mechan-*

ics, the infinitely many states of a quantum system are the coordinates of a point in a Hilbert space, and the physical quantities of the system (e.g., position and velocity) are represented by certain functionals—or, in the usual terminology, *operators.* The physics of quantum mechanics is thus reduced to the mathematics of particular (linear hermitian) operators on a Hilbert space. For instance, the famous *Heisenberg uncertainty principle*, according to which it is impossible to specify simultaneously the position and the velocity of a particle as accurately as we wish, translates into the noncommutativity of the corresponding operators.

Stimulated by these applications to physics, the study of operators that represent the physical quantities of a system became an important branch of modern mathematics, under the rubric *von Neumann algebras of operators.* These algebras can be factorized in various ways, for example into two sets of operators, where the elements of the first set commute with those of the second. Besides these factors, known as type I, there are two other types: II and III. A complete classification of type III factors has been given by Alain Connes, who received for this result a Fields Medal in 1983. From a study of type II factors, Vaugham Jones derived his invariants for knots, to which we shall return later, and for this work he was awarded a Fields Medal in 1990.

As for the Banach spaces, their theory soon ran into a long series of difficult and apparently insurmountable problems, which provoked their temporary decline. The theory was reborn in the 1950s, when thanks to the new methodology introduced by the members of the French school, from Laurent Schwartz to Alexandre Grothendieck, Field medalists in 1950 and 1966, respectively, many classical problems could finally be solved. The field is now experiencing a third rejuvenation, as indicates the awarding of Fields Medals in 1994 and 1998 to Jean Bourgain and William Gowers. The former has determined the largest section of a Banach space that resembles the

Hilbert space. The latter has shown that the only Banach space with a lot of symmetry (i.e., isomorphic to each of its subspaces) is the Hilbert space, and that there exist Banach spaces with little symmetry (i.e., nonisomorphic to any of its proper subspaces).

3.5. Probability Theory: Kolmogorov's Axiomatization (1933)

The first problems of a probabilistic nature arose in connection with games of chance, in particular those involving dice. A nontrivial one is quoted in Luca Pacioli's *Summa*, dating from 1494. If in a given game the first of two players to obtain n points wins, but the game stops when they have reached p and q points, respectively, how are their bets to be divided?

The problem was examined by Cardano in his 1526 *Liber de ludo aleae*, in which, among others, the multiplication rule is explicitly stated: the probability of two independent events occurring jointly equals the product of their individual probabilities.

The correspondence about the problem between Blaise Pascal and Pierre de Fermat in 1645 marks the official birth of probability theory. The solution required certain properties of the so-called Pascal's triangle, that is, the coefficients of the binomial expansion. One must calculate the probability of a player to win all the remaining points, all these minus one, all minus two, and so on, until reaching the minimum score, which, added to the points the player already has, allows him or her to win the game.

In 1656 Christian Huygens published Pascal's solution and introduced the concept of *expected value*, or how much one can expect to win, on average, by playing a game a large number of times, which corresponds to how much one should be willing to pay to play the game. Huygens considered that a measure of the expected gain in a given situation was the product of the possible gain times the probability of obtaining it; and a measure of the total gain was the sum of the expected gains for each possible situation.

A paradox involving the notion of expected value was discovered by Daniel Bernoulli in 1725. If a casino was willing to pay 2^n dollars to a player who obtains "heads" for the first time at the nth flip of a coin, how much should the player be willing to pay to play the game?

Since for each successive trial the gain doubles but the probability of winning is halved, the expected gain after n trials is the same for all n; and the total expected gain is therefore infinite. The player should then be willing to bet all he or she has. This contradicts the obvious observation that, the more the player pays to play, the smaller the probability of winning more than what he or she has paid.

The solution of the paradox proposed by Bernoulli rests on the fact that the value of money is not absolute but depends on how much one has; a given sum is worth a lot to a person who has little, but is worth little for someone who has a lot. To calculate the expected gain, one must then multiply the probability, not by the actual gain, but by its value for that particular player, that is, by its *utility*. Assuming, for example, that the utility decreases logarithmically, the total gain is no longer infinite but becomes very small, and the paradox disappears.

The first book on probability theory was the *Ars conjectandi* by Jacques Bernoulli, Daniel's uncle, published in 1713. The *law of large numbers* appears in it: if an event occurs m times in n trials, as the number of trials increases, the ratio m/n keeps getting closer to the probability of the event. This law allows us in principle to calculate probabilities a posteriori, when it is not possible to count a priori the number of favorable and of possible outcomes.

In practice, however, there remains the problem of statistically inferring the probability of an event from the fact that it has occurred m times in n trials. In 1761 Thomas Bayes worked on the problem, and his solution required the statement of *Bayes's law*: the probability of two events to occur at the same time is the product of the probability of one of them occurring (without conditions) times the probability of the other occurring conditionally to the occurrence of the first.

In 1777 Georges Louis Leclerc, count of Buffon, considered the following *problem of the needle*: given a lined sheet of paper, what is the probability that a needle with length one-half the distance between the lines will fall on one of them, after being dropped over the sheet in a random manner? Since the way the needle falls depends on its angle of inclination with respect to the lines, one would expect the answer to depend in one way or another on π. Sure enough, Buffon showed that the probability is $1/\pi$. By the law of large numbers, we can then approximate the value of π by dropping the needle a large number of times. This was the first application of what is today called the *Monte Carlo method*: to calculate the value of some constant, first show that it is equal to the theoretical probability of a certain event, and then perform empirically a large number of practical simulations of the event.

In 1809 Gauss found the famous bell-shaped curve, with equation e^{-x2} , which describes the probability distribution of average errors in repeated observations (fig. 3.8). The curve is symmetrical, because it is equally likely that the error be larger or smaller than the actual value, and the curve decreases toward zero in each direction, because the probability of very large errors is very small. Of course, there are many curves with those properties. As is clear from the exponent, Gauss obtained his curve by the *method of least squares*: the best approximation to a set of observations is that for which the sum of the squares of the individual errors is a minimum.

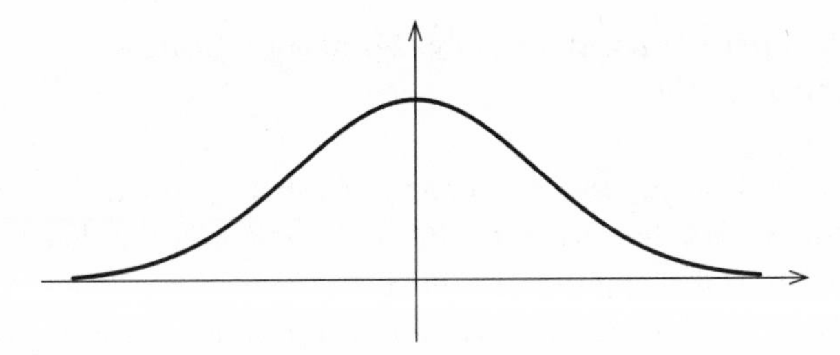

Fig. 3.8 Gauss curve.

All these developments came together in 1812 in the treatise *Théorie analytique des probabilités* by Pierre Simon de Laplace. Laplace systematized the subject by defining the probability of an event as the ratio between the favorable and possible outcomes; by showing that the area under the Gauss curve is $\sqrt{\pi}$; and by considering all kinds of applications to the natural and social sciences.

If probability had thus achieved maturity, an abstract definition of it was still missing. The request for such a definition was part of *Hilbert's sixth problem*, and, in 1931, Andrei Kolmogorov, 1980 Wolf Prize winner, obliged, unexpectedly using for this purpose the concept of a Lebesgue measure.

Kolmogorov's idea was to define axiomatically not just the probability of a single event but also that of a set of events. To each set of events is assigned a number between 0 and 1 (its probability) satisfying the following properties: the empty set of events has probability 0, and the set of all possible events has probability 1; a set of events obtained by "adding" together a countable number of (nonoverlapping) sets of events has probability equal to the sum of their probabilities (*countable additivity*).

If there is only a finite number (n, say) of single events that are all equally likely, the above definition assigns probability 1 to the entire set; thus, the sum of the individual probabilities is 1, and since these are all the same, each single event must have probability $1/n$.

3.6. Optimization Theory: Dantzig's Simplex Method (1947)

Some contrasting but convergent factors led, in the first half of the twentieth century, to the development of the theory of economic programming. In the Soviet Union, planning was a theoretical consequence of the birth of communism and found its concrete expression in the practice of the quinquennial plans. In the United States, on the other hand, planning became a practical necessity for the development of capitalism, and it gave birth to the theory of operations research for the management of large companies.

But it was above all during the war effort of the second world conflict that problems of a technical nature arose. The search for solutions would lead to the building of the computer, and to the technique known as *linear programming* for finding the best allocation of a certain number of resources according to a given optimization criterion. The qualifier "linear" refers to the essential characteristic of the problem, which is to impose constraints among the resources expressed in the form of linear inequalities, and to set an optimization criterion expressed as a linear function.

When there are only two resources, these can be represented as points in the plane, and each inequality then determines a half-plane. Excluding the extreme cases of an empty or unbounded intersection, the set of inequalities determines a convex polygon whose points constitute the solutions to the problem. Among these, the best one must be selected, according to the given optimization criterion. To find this optimal solution, it is not necessary to examine all possible solutions and compare the values of the optimization criterion; it is enough to consider the vertices (since the polygon is convex, each internal point is in a segment whose endpoints belong to the perimeter; given that the criterion is linear, it attains its maximum value over the segment at one of the endpoints, that is, at a point of the perimeter, and the maximum value of the criterion over the entire polygon occurs at one of its vertices).

When there is a large number of resources and constraints, the polygon becomes a polyhedron in a multidimensional space, and even limiting the search to the sole vertices may present insurmountable difficulties. The classical solution to the problem was the *simplex method*, developed in the 1940s by George Dantzig, Leonid Kantorovich, and Tjalling Koopmans, and for which the last two obtained the Nobel Prize in economics in 1975.

The general idea of the method, which due to its efficiency became one of the most widely used algorithms in the history of applied mathematics, is to start at some particular vertex of the polyhedron, and move over to an adjacent vertex with a better value for the optimization criterion. Proceeding in this way, a locally optimal value is reached. Since the polyhedron is convex, a locally optimal value is also globally optimal, and so the method always produces the best solution.

One of the hypotheses underlying linear programming is that the values of the resources may be fractions. The reason for this is that the vertices of the polyhedron determined by the inequalities are found by solving systems of linear equations, whose solutions are generally not integers. If in a given application the resources can only take integer values, as it is often the case in practice, it is not enough to proceed with the optimization assuming the resources could be fractions, and then rounding off the solutions at the end; in fact, minor changes in value might produce a different optimal vertex. It has therefore been necessary to extend linear programming by developing specific techniques to solve *integer programming* problems.

Another necessary extension involved methods for the solution of nonlinear problems. In this case the simplex method does not work for a different reason: without linearity (and hence without convexity) a local optimal point may no longer be a global one. The necessary conditions for the existence of an optimum, which are the basis for the many algorithms for solving nonlinear programming, have been provided by Harold Kuhn and Albert Tucker in 1950, in a seminal paper that also gave the name to the subject.

3.7. General Equilibrium Theory: The Arrow-Debreu Existence Theorem (1954)

In 1776, the year of the American bourgeois revolution, the Scottish economist Adam Smith published the treaty *Inquiry into the Nature and Causes of the Wealth of Nations.* To justify the policy of laissez-faire, he introduced the rhetorical fiction of an "invisible hand" that would supposedly guide the individualistic behavior of the economic agents toward goals they had not anticipated, and that would turn out to be socially useful. However, the argument was based on a vicious circle, summed up in the optimistic maxim "all is right with the world."

The first attempts to found a science based on Smith's economic philosophy had to wait until the nineteenth century. In 1838 Antoine-Augustin Cournot resorted to the tools of infinitesimal calculus, from functions to derivatives, to describe the fundamental concepts of economics. In 1874 Léon Walras established a parallel between economics and mechanics, in which the laws of the market and market equilibrium were seen as the economic counterparts of the law of gravitation and mechanical equilibrium. The analogy was extended at the end of the century by Vilfredo Pareto, who considered the economic agents as the equivalent of physical particles.

In particular, Walras stated a theory that replaced Smith's undefinable invisible hand by the interaction between supply and demand, and conjectured that the development of the market naturally tends toward their equilibrium. In mathematical terms, one must express supply and demand for each product as a function of the price and the availabilty of all the goods, and then impose the condition that the difference between demand and supply should always be zero. This being the case, for each merchandise the amount produced would exactly match the amount that is sold. The problems to be solved are, first of all, the *existence* and *uniqueness* of an equilibrium, that is, of a system of prices that would satisfy all the equations; then, the automatic *convergence* of the system toward the equilibrium on the basis of the law of supply and demand, and according to which prices rise as demand increases, and fall when demand decreases; finally, the *stability* of the equilibrium, in the sense that even if the system is momentarily perturbed, it will tend to return to its equilibrium state.

Of course, it all depends on the particular form of the functions expressing supply and demand, on the one hand, and on the laws governing them, on the other. Walras defined a system of nonlinear equations, and derived the existence of a solution from the fact—certainly not sufficient—that there are as many equations as unknowns. In 1933 the economist Karl Schlesinger and the mathematician Abraham Wald proposed a different system, and gave for the first time a formal proof of the existence of equilibria.

In 1938 John von Neumann came up with two novel ideas. First, he reformulated the problem in terms not of equations, as had until then been done, but of inequations. This paved the way for a formulation analogous to that of optimization problems, and for the solution of the linear ones by Dantzig's simplex method. In addition, von Neumann showed the existence of an equilibrium for a particular system by reducing it to a minimax problem, and then using a version of Brouwer's fixed point theorem. Von Neumann's ideas on game theory and equilibrium theory reached their final form in 1944, in the already mentioned book *The Theory of Games and Economic Behavior.*

The essential feature of von Neumann's proof of the existence of an equilibrium was to shift the attention from the techniques of classical differential calculus to topology, and hence from dynamical to static systems. Through this new approach, and in particular by using an extension of Brouwer's fixed-point theorem proved in 1941 by Kakutani, Kenneth Arrow and Gerard Debreu finally succeeded in 1954 in proving the existence of an equilibrium for the Walras equations, where the law of supply and demand is stated as follows: the rate of change in the price of each of the goods—and hence its derivative with respect to time—is proportional to the excess demand (i.e., the difference between demand and supply for that particular product). This work earned their authors the Nobel Prize in economics, in 1972 (Arrow) and 1983 (Debreu).

The use of Brouwer's fixed-point theorem permitted Arrow and Debreu to get around the difficulties that arise in the study of economics through dynamical systems, which in the 1950s had not yet been sufficiently developed. These, however, made a comeback in the second half of the century, helped by the new possibilities offered by computer simulations. In 1982 Stephen Smale, awarded the Fields Medal in 1966 for other work that we shall mention later, closed the historical circle by proving the Arrow-Debreu theorem with the methods originally intended by Walras, and without using any fixed-point theorems.

To be sure, before being able to derive from the equilibrium theorem any political conclusions, which would vindicate in some sense the liberalism à la Adam Smith, it would be necessary to prove it in a more general setting than the simplified version of Arrow and Debreu; in particular, in a situation in which the various markets interact among themselves, and the variation in the price of each product depends (e.g., linearly) on the excess demand of all goods, and not just of the product in question.

Unfortunately for capitalism, in these more general conditions a market tends by itself toward equilibrium only in the rather special case when there are only two goods. In 1960 Herbert Scarf showed that it only takes three products for the system to be (potentially) globally instable, and not at all guided by the hypothetical invisible hand. And in 1972 Hugo Sonnenschein proved that the global excess demand of a market may take on the values of any given polynomial. Thus, the market may not automatically tend toward the equilibrium points, or come back to them after having been disrupted.

If we may draw a political conclusion from these mathematical developments, it is that the law of the market does not appear at all sufficient to drive it toward equilibrium, and that only some sort of planning can do it—Adam Smith, and his modern disciples, from Margaret Thatcher to Ronald Reagan, notwithstanding.

3.8. The Theory of Formal Languages: Chomsky's Classification (1957)

One of the most significant turning points in modern linguistics was the series of lectures given by the Swiss linguist Ferdinand de Saussure between 1906 and 1911, and published posthumously in 1916 under the title *Cours de linguistique générale*. These lecture notes contain the basis for a structural approach to natural languages, as opposed to the historical, philological, and comparative studies that had been the norm until then. Saussure regarded language as a two-part system: on the one hand, a fixed, collective, and immutable structure of rules for symbol manipulation (oral or written); on the other, a variable, individual, and creative use of the structure for the expression of meaning.

Saussure's ideas pointed to the possibility of a mathematical study of the structural aspects of linguistics, and more generally of the social sciences. His ideas did in fact anticipate and inspire *structuralism*, whose goal was the search for underlying structures in the manifestations of human experience, and which found its concrete expression in the anthropology of Claude Lévi-Strauss, the psychoanalysis of Jacques Lacan, and the psychology of Jean Piaget.

For its part, the axiomatic and formalistic conception of mathematics also led, naturally and independently, to ideas that run parallel to those of Saussure's, namely, that it is possible to reduce linguistic activity to the generation of strings of symbols according to some formal rules, and that symbols are linked to meaning in a conventional and arbitrary manner.

It is not by chance if the first formulation of abstract and formal rules for the description of linguistic structures goes back to the mathematician Axel Thue. In 1914 he expressed these rules in terms of *grammatical productions* of the type

$$x \rightarrow y,$$

which means that every occurrence of x in a word may be replaced by y. Thue defined a *grammar* as a set of productions of the above form and posed the so-called *word problem*: decide whether any two given words may be transformed into one another by using the productions in the grammar.

In 1921 Emil Post arrived, independently, to a similar formulation, and he proved a surprising result that today may be stated as follows: the languages admitting a Thue grammar are precisely those that can be generated by means of any one of the usual programming languages for a computer. In other words, simple grammatical productions are enough to describe all that the most sophisticated computer programs can do, in particular all possible types of formal or mechanical languages.

There remained to treat the case of human languages. This was the task undertaken in 1957 by the linguist Noam Chomsky, who in *Syntactic Structures* took the first steps of a program that should have led to a complete description of a Thue grammar for English. But this program was never achieved, and the difficulties encountered seem to indicate a structural shortcoming of the purely mathematical approach for the study of natural languages.

Chomsky's work produced nevertheless a fundamental result in the theory of formal languages, namely, a classification based on the type of grammatical productions allowed. And since the same types of languages later turned out to be characterized also by the type of computer capable of generating them, his result was the starting point of the theory of formal languages for computers, that is, of computer linguistics.

Chomsky's classification identifies four types of languages: *universal*, *context-sensitive*, *context-free*, and *regular*. Roughly speaking, in the first type there are no restrictions on the kind of grammatical productions, and therefore the substitution of any part of a word by another word is allowed. In the second type, substitutions of parts of a word by another word are permitted only in particular contexts, which are specified in the productions. In the third type, only a single letter may be substituted by a part of a word; and, finally, in the languages of the fourth type only a single letter may be replaced with another single letter.

For each of the four language types in the classification there is a corresponding type of computer or automata that can generate it: *universal*, *linear*, *push-down*, and *finite*. Basically, in the first type there are no restrictions on the amount of computer memory or storage available; in the second, the machine cannot use an amount of memory that is larger than the input; in the third type the computer can only store data as in the piles of trays in a cafeteria, where the first trays to have been placed are at the bottom, and will therefore be the last to be removed, and vice versa. Finally, in the fourth type the computer may read data, but cannot store it.

From a linguistic point of view the most interesting grammars are the context-sensitive, but for the computer scientists the most useful turned out to be the context-free and the regular grammars, and their theory is nowadays a well-developed branch of theoretical computer science.

As for pure mathematics, the most interesting applications of formal linguistics were related to the word problem proposed by Thue. Many algebraic structures naturally present themselves in the form of productions, for example groups and semigroups (the latter are a weaker version of groups in which the existence of an inverse for each element is not required).

Post, in 1944, and Anatoly Markov, in 1947, showed that there is no algorithm for deciding the word problem for semigroups. This was the first example of an undecidable problem that was not artificial, and it served to illustrate that the limitations of formal systems discovered by Gödel, Church, and Turing concern not only the theoretical foundations of mathematics but also its practice.

In 1955 Pavel Novikov, and in 1959 William Boone, proved that the more difficult word problem for groups was also undecidable. Because of its connection with the fundamental groups in algebraic topology that we shall discuss below, this result was later used to establish the undecidability of many topological problems, for instance, whether a surface is connected, or two surfaces are topologically equivalent.

3.9. Dynamical Systems Theory: The KAM Theorem (1962)

The mathematical study of the motion of bodies was made theoretically possible by Newton's discovery, in 1664–66, of the infinitesimal calculus, on the one hand, and of the three *laws of motion*, on the other, namely, the principle of inertia, the famous equation $F = ma$, and the principle of action and reaction. In the particular case of the celestial bodies, the force in action is specified by the law of universal gravitation: the attraction exerted by a body is directly proportional to its mass and inversely proportional to the square of its distance.

For example, in the first book of the *Principia*, Newton showed that the motion of a planet around the Sun obeys the three laws that Kepler stated in 1618: the orbit is an ellipse, with the Sun at one of its foci; equal areas are swept in equal times; and the square of the planetary year is (approximately) proportional to the cube of the average distance from the planet to the Sun.

In practice however, the planets are not only subject to the Sun's gravitational force but they also attract each other. As a consequence of this, their orbits are not perfect ellipses, nor are they necessarily closed. What is more, besides the Sun and the nine planets there are other celestial objects in the solar system (such as satellites, comets, and asteroids). The problem of its motion is therefore far from obvious.

The case of the Sun and a single planet is very special, because one of the bodies has a negligible mass compared to the other. We may then assume that the large one is at rest and the other revolves around it. Newton showed that the solution is nevertheless similar in the general case: both bodies move along elliptic orbits, with the barycenter of the system at one common focus.

The case of two bodies being solved, the next step becomes finding a solution for the *three-body problem*. Particularly interesting examples are the Sun, the Earth, and the Moon, or the Sun and two planets. Approximate solutions may be obtained by first solving the problem for two bodies, and then modifying the solution by taking into account the influence of the third one. Such was the method used first by Newton in 1687 to calculate the effect of the Sun on the Moon's motion around the Earth, and later by Euler in 1748 to calculate the perturbation caused by Jupiter and Saturn to each other's motion around the Sun.

Exact solutions of some special cases of the three-body problem were found by Joseph Louis Lagrange in 1772. He proved, for instance, that it is possible to have three bodies moving on three elliptic orbits, with the barycenter of the system at one common focus. Or that, if the bodies are located at the vertices of an equilateral triangle, the triangle rotates around the barycenter of the system, and the bodies remain fixed to the vertices. This later situation, as it was discovered in 1906, corresponds to the system formed by the Sun, Jupiter, and the asteroid Achilles.

Between 1799 and 1825 appeared the five volumes of Laplace's *Celestial Mechanics*, the culmination of one and a half centuries of discoveries. In particular, Laplace could claim that the evolution of the universe, past and future, could have been exactly calculated if the position and velocity of each body at one single instant were known.

Despite Laplace's optimism, two fundamental problems remained open. One was the exact solution of the body problem in the general case of three or more bodies; and the other the question of the stability of the solutions. For instance, would small perturbations of a planet's motion only cause small variations in its orbit, or could they instead send the planet wandering into the vastness of space? Another particular question concerned the cumulative effect of the mutual perturbations among the various planets. Would it be sufficient to throw one of them out of its orbit, and eventually out of the solar system, or are the planetary orbits expected to remain essentially in the present configuration?

The problem of the stability of the solar system eventually came to the attention of King Oscar II of Sweden, who added it to the list of problems whose solutions were to be rewarded with a special prize created in 1885 to "honor his sixtieth birthday and to prove his interest in the advancement of the mathematical sciences."

The prize was awarded to Poincaré in 1889 for his memoir, *On the Three-Body Problem and the Equations of Dynamics.* He had not managed to decide whether the solar system is stable or not, but contributed what amounted to a qualitative leap in the study of dynamical systems. He introduced what he called the new methods in celestial mechanics, which was precisely the title (*Méthodes nouvelles de la mécanique céleste*) of a trilogy he published between 1892 and 1899. These new methods included, in particular, the topological study of nonlinear differential equations, that until then had been put aside due to their difficulty.

The distinction between stable and unstable orbits is related, in an unexpected way, to problems in number theory. For instance, the ratio between the planetary years of Jupiter and Saturn is 5 to 2—a rational number. Therefore, every ten years both planets find themselves in the same positions, and so their mutual perturbations can in principle be amplified by a resonance phenomenon, and may end up producing a destabilizing effect.

The mathematical translation of the difficulty is the so-called *small divisors problem.* When the mutual perturbations of the two planets are expressed as an infinite sum (called a *Fourier sum*), the rational ratio 5/2 forces many coefficients of the terms of the sum to have small divisors, and therefore to be very large; this has the effect of making the sum tend to infinity. The 270-page work which earned Poincaré the "Oscar prize" appeared to indicate that such sums were in fact infinite, and thus that the orbits were unstable.

The stability problem was taken up again by Kolmogorov in 1954. He outlined a solution method, and his program was carried out with success by Vladimir Arnol'd and Jürgen Moser in 1962, in a piece of work that is globally called the *KAM theorem*, after its authors' initials. The solution shows that, for small perturbations, most of the orbits are stable, and even if they are not periodic, they remain close to the periodic orbits of the nonperturbed system, and for this reason are called *quasi-periodic.*

The mathematical essence of the KAM theorem is that the small divisors problem actually arises in the case of rational periods, or of periods that can be well approximated by rationals (i.e., through fractions with relatively small denominators), but otherwise it does not arise. Since the majority of real numbers are not well approximated by rationals, the problem does not come up in the majority of cases.

The interest provoked by the KAM theorem and the problems related to it is significant. If we consider it from the angle of pure mathematics, the original result paid off in the form of Wolf Prizes for Kolmogorov (in 1980) and Moser (in 1994–95), and for a recent generalization Jean Christophe Yoccoz received a Fields Medal in 1994. In the complementary, applied mathematics direction, the theoretical stability of planetary orbits in the solar system translates into the concrete stability of elementary particle orbits in the particle accelerators, a stability that is crucial if the particles are not to lose their energy in collisions against the accelerator walls. The relevance of the theorem stems from the fact that the number of particle orbits in an experiment is so large that it is of equal order of magnitude as the number of planetary orbits in the entire life of the solar system.

3.10. Knot Theory: Jones Invariants (1984)

According to the legend, in Gordium, an ancient city of Phrygia (today Turkey) the pole of King Midas's wagon was attached to its yoke by a knot so tight and entwined that it was said that whoever succeeded in untying it should reign over the entire world. Alexander the Great reached Gordium in A.D.

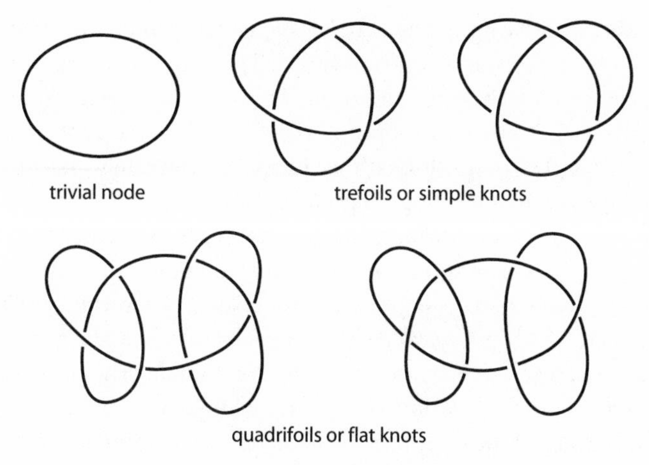

Fig. 3.9 Knots.

333, and after several unsuccessful attempts he cut the knot with a stroke of his sword. Of course, the problem was not solved, for the solution requires that a knot must be deformed without cutting it up, that is, in a topological way.

In 1848 Johann Listing, a student of Gauss, coined the name *topology* and published the first book on the subject. A good part of the book was devoted to the study of knots, that is, of closed curves in space (fig. 3.9). Knots can thus be considered one-dimensional surfaces, and so it is natural to look at them as topological objects, that is, as made of very thin rubber strings with their ends glued together, and to try to classify them just as Riemann, Möbius, and Klein did for 2-dimensional surfaces, and Thurston for those in 3 dimensions.

There is in principle a connection between knot theory and the theory of surfaces. Given a knot, we can envisage its support not as an abstract mathematical curve, whose section is just a point, but as a solid physical tube, whose section is a circle. Considering the 2-dimensional surface of the tube does not take us very far, because it is always topologically equivalent to a torus, for any knot. But we can consider the 3-dimensional surface consisting of the tube's cast, namely, the entire space minus the tube, its interior included. The structure of the knot then becomes the structure of the holes of this surface, and to study the latter we can apply all the classical topological tools. In particular, in 1978 Geoffrey Hemion completely classified all knots from a topological point of view.

This is, however, a very indirect approach, so knot theory has sought to assign *invariants* directly to knots. These invariants, as their name suggests, do not change when the knot is subjected to topological deformations—i.e., when the rubber band making up the knot is stretched or pulled without breaking it. Many of these invariants can be implicitly derived from the associated surface, but the problem is to define some explicit ones that may be obtained directly from the image of the knot itself.

The simplest invariant one can think of is the *number* of times the string intersects itself when it lays flat on a plane. Obviously, deforming the knot may change this number, for example, by twisting the string and thus adding artificial intersections. To really have an invariant one must then take the smallest number of intersections needed to represent the given knot. But this approach renders the invariant almost useless, for to calculate it one should already know the type of knot under consideration.

In 1910 Max Dehn introduced an algebraic description of knots, which allowed him to prove that there are different types of knots. In other words, that not every knot may be untied, and thus be reduced to the trivial knot (i.e., a circle) by appropriate deformations and without breaking it. This fact is intuitively clear already for the trefoil (or simple knot), but the question was to prove it mathematically.

In 1928 James Alexander defined an invariant that is a polynomial, and which, besides the mere intersections, takes into account also the way in which these occur (the variable of the polynomial represents the meridian of the knot). When two knots are added, their Alexander polynomials are multiplied. Since the trefoil has polynomial $x^2 - x + 1$, and the quadrifoil (or flat knot) is the sum of two trefoils, its polynomial is

$$(x^2 - x + 1)^2 = x^4 - 2x^3 + 3x^2 - 2x + 1$$

It can be shown that if two knots have different polynomials, they must be different. It follows from this that the trefoil and the quadrifoil cannot be obtained one from the other by deformations; and that there are infinitely many different knots, because each (symmetric) polynomial is the polynomial of some knot. However, two knots can be different and still have the same polynomial, as in the case of the right-handed and left-handed trefoil.

In 1984 Vaughan Jones defined as an invariant a new type of "polynomial" (in quotation marks, because the exponent of the variable can also be a negative integer) which takes into account also the direction of the intersections, and thus it is able to distinguish between the two trefoils. Their polynomials are, respectively,

$$-x^4 + x^3 + x \quad \text{and} \quad -\frac{1}{x^4} + \frac{1}{x^3} + \frac{1}{x}.$$

Jones arrived at his "polynomials" indirectly, through the study of the von Neumann algébras, and he later discovered a further unexpected connection with statistical mechanics. For these results, as well as for the fruitful consequences of his invariants, Jones was awarded the Fields Medal in 1990.

Despite these developments, a complete classification of knots has not yet been found. In particular, a complete invariant is still missing, one that would permit to distinguish among them all knots that are actually different (the best invariant to date is due to Maxim Kontsevich, for which he received the Fields Medal in 1998). Even in its present incomplete state, the applications of knot theory are extremely significant.

Let us begin with physics. In 1867 Lord Kelvin proposed a theory according to which the atoms were knots in the ether, called *vortex atoms*, similar to whirls of smoke in the air. The strange idea was based on a theorem due to Hermann Helmholtz and stating that in a perfect fluid, once a vortex occurs, it stays forever. Kelvin was inspired by experiments carried out by Peter Tait with smoke rings, which bounced elastically and exhibited some interesting vibration patterns. The advantage of such a theory was that the knots were kept together by purely topological bonds, without the intervention of any specific atomic forces. Kevin's proposal stimulated a ten-year study of knots by Tait, and produced a rather complete and accurate table of the knots having up to ten intersections. But the theory was abandoned when Bohr's model, which represented the atom as a miniature solar system, was adopted.

Knots are today fashionable thanks to *string theory.* These strings could be the ultimate constituents of matter, and the elementary particles would be vibration patterns of strings in multidimensional spaces. In fact, there are several string theories. In the simplest one, the strings are open and unidimensional, like little pieces of string with quarks attached to the ends, but in other theories the strings may be closed, like the knots we have discussed. In most recent theories, one-dimensional strings are replaced with multidimensional *membranes* that can be open or closed.

Many of the mathematical ideas behind string theory have their roots in Edward Witten's pyrotechnical works, which have profoundly influenced mathematics in recent years and won him the Fields Medal in 1990. Witten found unexpected relations between string theory and the most diverse areas of mathematics. For example, the Fischer-Griess monster in group theory, Jones polynomials in knot theory, and Donaldson exotic spaces in topology all turned out to be different aspects of certain topological-quantum field theories, in 2, 3, and 4 dimensions, respectively.

From this perspective some mysterious symmetries of these objects may be explained and their scope considerably extended. For instance, it is precisely by using string theory that Maxim Kontsevich and Richard Borcherds obtained the results that won them the Fields Medal in 1998. Kontsevich was able to generalize Jones polynomials and obtain new invariants, not only for knots but also for tridimensional surfaces (Jones polynomials turned out to be Feynman integrals calculated over a particular surface, whose definition comes from string theory). Borcherds was able to solve the *moonlight conjecture*, proposed by John Conway and Simon Norton in 1979, which relates the Fischer-Griess monster with the theory of elliptic functions introduced in 1827 by Niels Abel and Carl Jacobi (it was found that the monster is the automorphism group of a particular algebra, whose axioms are obtained from string theory).

In the recent versions of string theory, the previously mentioned Calabi-Yau manifolds play a fundamental role. In a first phase, known as *supersymmetry*, it was discovered that the imposition of strong invariance conditions to string theory called precisely for a model involving a Calabi-Yau manifold. The three complex dimensions of the manifold correspond to six real dimensions, which added to the four dimensions of space-time produce a total of ten dimensions. In a second phase, the so-called *mirror symmetry*, it was found that the physical theory could be modeled using two different Calabi-Yau manifolds, and that some difficult calculations in one of the manifolds turned out to be easy in the other, and vice versa. And so, with each foot in a different shoe, it was possible to take some essential steps forward in the search for a *theory of everything* capable of encompassing all of modern physics.

A different type of application of knot theory is the study of the structure of DNA, consisting in a long filament of genes folded over on itself, a chain almost one meter long and dwelling in the nucleus of a cell whose diameter is 5 millionths of a meter (imagine a 200-kilometer-long thread packed inside a soccer ball). When DNA replicates, it separates into two identical copies. The question is to understand how this can take place in an efficient way, given that the analogous separation of the threads making up a string results in complicated knots and entanglements. While Alexander invariants proved inadequate to deal with DNA folds, Jones invariants, on the other hand, have already produced interesting results in this field.

CHAPTER 4 MATHEMATICS AND THE COMPUTER

Computers are changing daily life in a fundamental way, not only the life of the ordinary person, but also that of the mathematician. As is often the case with technology, many changes are for the worse, and the mathematical applications of the computer are no exception. Such is, for example, the case when the computer is used as an *idiot savant*, in the anxious and futile search for ever larger prime numbers. The record holder at the end of the twentieth century was $2^{6,972,593} - 1$, a number that is approximately 2 million digits long.

The inherent dangers of the mindless use of the computer are admirably illustrated by the following story, which shows how an indiscriminate reliance on the power of the machine may hinder rather than stimulate mathematical thought. In 1640, Fermat had conjectured that all numbers of the form $2^{2^n} + 1$ are prime, based on the fact that such is the case for values of n from 0 to 4, which result in the numbers 3, 5, 17, 257, and 65,537, all of which are indeed prime. Today, a computer can easily check by brute force that Fermat's conjecture is already false for $n = 5$, since

$$2^{2^5} + 1 = 2^{32} + 1 = 4{,}294{,}967{,}297 = 641 \times 6{,}700{,}417.$$

But a systematic manual search for the possible factors was, and still remains, impossible. In 1736 Leonhard Euler avoided such a search by showing, with a clever and drastic reduction, that it was enough to consider factors of the form $64k + 1$. The factor 641 can then be found at the tenth trial ($k = 10$). The unavailability of a computer thus forced Euler to shift the problem from mere accounting to higher mathematics, and to solve one of Fermat's curious problems through the use of one of his own surprising theorems. For the record, no other Fermat prime numbers are known, and in 1990 the combined power of one thousand computers made it possible to emulate, for $n = 9$, what Euler had done by hand for $n = 5$, without, however, producing any interesting mathematical results.

Both the particularity of this episode, and the generality of Wittgenstein's *Philosophical Investigations* motto, warn us that "all progress looks bigger than it is." In other words, the consequences of the use of the computer, in mathematics and elsewhere, must not be thoughtlessly exaggerated, as it is too often done in the popular press, but must rather be subjected to a critical analysis that would permit us to distinguish the silhouette of substantial progress against the background of superficial developments.

First of all, we must acknowledge that the possible impact of computers on mathematics would be no more than a return of favors. For if it is often true that scientific theories follow technological advances, in this particular case the opposite took place. Indeed, the construction of the first electronic computers was the culmination of a mathematical development that lasted a full century, and which consisted of three main stages.

The first fundamental idea was introduced in 1854 by George Boole in *The Laws of Thought*. In his famous treatise, Boole presented the algebraic formulation of the semantic behavior of the simplest linguistic particles, such as conjunction and negation, which is today known as *Boolean algebra*. The mathematical treatment of the laws governing the thinking process was carried on by Frege and Russell, who extended it sucessfully to the whole of logic. And postwar artificial intelligence sought, although with limited success so far, to further extend the formalization of thought even beyond the logical and rational domain.

The second significant idea was due to Alan Turing. Starting from the logical calculus of Frege and Russell, he demonstrated in 1936 that there cannot exist a procedure for deciding whether any given formula of such calculus is valid or not. In other words, it is impossible to mechanize the semantics of logical reasoning, in a way similar to what had been achieved for its syntax. To prove his impossibility result, Turing introduced the notion of an abstract machine capable of executing all possible formal calculations and showed that such a machine could not solve the decision problem. Turing's machine may be described in modern terms simply by saying that it was the theoretical blueprint for a universal computer.

To physically implement such a machine required one more idea, which was born out of the collaboration between a neurophysiologist and a mathematician: Warren McCulloch and Walter Pitts. Since what was missing was a "brain" to guide Turing's machine through the execution of its calculations, they proposed in 1943 an abstract model of a nervous system, based on a simplification of the one humans possess. They also showed that such an artificial system could be built by means of electric wires whose connections play the role of neurons, and where the flow, or lack of it, of an electric current represents the presence or the absence of a synaptic response. What those neural networks could realize turned out to be precisely the Boolean algebras.

An electronic computer is nothing more than the practical implementation of Turing's machine combined with McCulloch's and Pitt's neural networks. The latter provide the former with a brain capable of making the most elementary logical decisions, thanks to which the machine can execute all possible mechanical calculations, excluding the decisions that require a logic of a higher order.

The developments we have merely touched upon played an influential role on the two projects leading to the construction of the first electronic computers: the Electronic Numerical Integrator and Calculator (ENIAC), directed in the United States by von Neumann, and the Automatic Computing Engine (ACE), under the supervision in Great Britain of Turing himself, both projects taking place around the 1950s. Thus, since the computer was the offspring of mathematical research undertaken in the first half of the century, it is not surprising that the machine would later exhibit traces of its genetic legacy.

The first mathematical application of the new device was of course the use of its computational power. In fact, its very conception had been stimulated precisely by the desire to automate the tremendous amount of calculations required by the war effort, which both Turing and von Neumann had experienced firsthand, the former while doing counterintelligence work, and the latter during the construction of the atomic bomb. This use of the computer to perform computations is still today its most widespread function, and it accounts for its very name.

The advantages of being able to carry out large numbers of calculations at high speed had no doubt an impact also in pure mathematics. The most famous case is certainly the 1976 interactive proof of the four-color theorem by Kenneth Appel and Wolfgang Haken, which necessitated help from the computer in the form of thousands of hours of machine time. Some twenty years later, in 1997, the first proof of a theorem entirely performed by a computer without human help took place. It concerned a conjecture formulated in 1933 by Herbert Robbins, who stated that a certain system of three equations was an axiomatization of the theory of Boolean algebras. Robbins's conjecture was proven true by a computer program written by William McCune and Larry Wos.

As was to be expected, it is rather in applied mathematics that the use of computers is having the most noticeable effects. For instance, until the second half of the twentieth century, the study of dynamical systems required a three-stage process: the translation of the system in mathematical terms, the explicit solution of the system, and the graphical description of the solution. Often the process could not proceed beyond the first stage, for the difficulty in obtaining the mathematical formulation prevented the solution of the system. This had resulted in the avoidance of complex systems and in a concentration of efforts on systems whose description was sufficiently simple for a solution to be workable. Even when solutions could be obtained, either explicitly or by some approximation process, their graphical representation might be impossible to achieve due to the extremely large number of calculations required.

The use of computers made it possible not only to solve the second problem but the first one as well. It is in fact often possible to avoid the search for an explicit solution of the mathematical description of a system and to obtain a graphical description of its behavior directly through a simulation. This allowed mathematicians to investigate a whole class of systems whose study had never been possible and resulted in the birth of what is now known as the *theory of chaos.* Its name notwithstanding, this theory consists in the study of systems that are not really chaotic, but so complex as to appear at first sight to be so.

The most famous metaphor regarding chaotic systems is "the butterfly effect": a butterfly flapping its wings in one continent may cause a tornado in another. One of the classical applications of computers, which began with von Neumann himself and was later continued by Edward Lorenz, is precisely the simulation of weather. This application made short-term weather forecasting possible and produced one of chaos's most

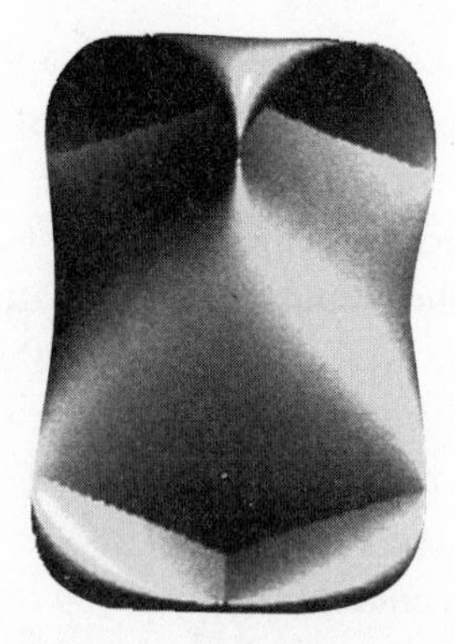

Fig. 4.1 Etruscan Venus.

popular images: a strange attractor in the form of, sure enough, butterfly wings.

Speaking of images, it is impossible not to mention the developments in computer graphics. Ever present in commercial applications, they are now playing an important role also in pure mathematics as a visual aid. The most typical example is the discovery of new surfaces that would have been hard to picture solely with the mind's eye. These are the minimal surfaces found in 1983 by David Hoffman and William Meeks, which we have already mentioned (fig. 2.5), and the so-called *Etruscan Venus*, discovered in 1988 by Donna Cox and George Francis (fig. 4.1).

The most widely known images, due also to their visual qualities, and which some would consider as the expression of a new art form, are those of *fractals*: the self-similar curves discovered at the beginning of the century as a mere curosity, temporarily abandoned due to the difficulty in their representation, and which made a startling comeback in the 1980s thanks to the work of Benoit Mandelbrot. The latter also discovered a kind of universal fractal, which when examined at a microscopic scale, reveals an inexhaustible supply of astonishing details, and whose images have become the symbol of the fruitful potential of a clever use of computers in mathematics.

Having thus introduced in general terms the problem of the mutual relationship between mathematics and computing, let us now examine in detail some of the most interesting applications of the computer to mathematical research, which we have already mentioned.

4.1. The Theory of Algorithms: Turing's Characterization (1936)

At the 1928 International Congress of Mathematicians held in Bologna, Hilbert proposed yet another of his famous problems, the so-called *Entscheidungsproblem*, or "decision problem": find an algorithm for deciding whether a given proposition is a logical consequence of other propositions.

Interest in this problem lies in the fact that the various branches of mathematics may be uniformly presented using axiom systems, from which the theorems follow as logical consequences. An algorithm such as the one sought by Hilbert would have thus allowed mathematicians to concentrate on the pleasant part of their work, that is, formulating axioms and stating interesting results, and to leave to the algorithm the laborious task of proving the results from the axioms.

However, the problem was more than wishful thinking. In 1922 Emil Post had already made substantial progress on it when he showed that propositional logic, the part of logic that deals with the linguistic elements known as *connectives* ("not," "and," "or," "if-then") admits in effect such an algorithm, the so-called *truth tables* method. Hilbert now proposed to extend this result to that part of logic dealing also with *quantifiers* ("none," "some," "all")—that is, predicate logic.

The problem was solved independently in 1936 by Alonzo Church in the United States and Alan Turing in England. The solution, as may be inferred from the fact that the search for proofs still constitutes the central part of doing mathematics, was a negative one: an algorithm of the kind Hilbert was looking for does not exist. But the proof of this fact presupposes a substantial progress: while a proof of the existence of an algorithm can be accomplished by merely exhibiting one with the desired properties, a proof on *non*existence requires the elimination of every possible algorithm, and hence the complete characterization of the very concept of algorithm.

The fact that such a vague and intuitive notion admits a precise and formal characterization was an amazing discovery. This was achieved through a series of attempts at definitions which, in the end, all turned out to be equivalent. But it was really Turing's approach that finally convinced mathematicians that the solution to the problem had been found. Nowadays his definition may be put in terms that appear almost trivial, by saying that an algorithm is a procedure that can be translated into a computer program, in any one of the so-called universal languages (for instance, the imperative language PASCAL, the functional one LISP, or the logical one PROLOG).

To be sure, in 1936 computers did not exist. Not only that, but their development was based precisely on Turing's introduction of the concept of a universal machine that can calculate every computable function by executing a program—more specifically, on the transition from dedicated machines able to perform only fixed calculations, such as computing machines, to universal machines capable of performing any executable calculation, such as computers.

Turing arrived at the negative solution of the *Entscheidungsproblem* by translating into the language of logic the so-called *halting problem*, that of deciding whether a given program will eventually terminate its computation and halt on given input data. This problem may be easily shown to be undecidable, in the sense that there is no program that can decide it, by using the classical diagonal method first introduced by Cantor in set theory and later employed by Russell in his paradox, and by Gödel in the proof of his incompleteness theorem. Such a method was therefore well known to Turing (as well as to Church, who solved the problem in a similar way, except that he used his equivalent definition of algorithm in terms of the lambda calculus).

The solution of the *Entscheidungsproblem* showed the way for undecidability proofs in a variety of fields, through suitable translations into the halting problem or other similar problems. From a mathematical point of view, the most interesting application of the method was the negative solution of *Hilbert's tenth problem*: find an algorithm that would decide whether a polynomial (in one or more variables) with integral coefficients (positive or negative) has integral zeros; or, in other words, whether the Diophantine equation obtained by equating the polynomial to zero has integral roots.

At the time of the 1900 congress, positive solutions to particular cases of Hilbert's tenth problem were known. For example, the Euclidean algorithm for finding the greatest common divisor can be used to treat the case of Diophantine equations of degree 1, since

$$a_1x_1 + \ldots + a_nx_n = b$$

has integral solutions if and only if the greatest common divisor of $a_1, \ldots, a_n$ divides b. And Gauss's quadratic reciprocity theory allows one to treat the case of Diophantine equations of degree 2.

A 1968 result, due to Alan Baker, which establishes effective upper bounds to solutions of polynomial equations of degree 3 or higher—and which earned its author the 1970 Fields Medal—can be used to treat the case of elliptic equations. This fact revealed a profound connection between Hilbert's tenth problem, Mordell's conjecture, and Fermat's theorem. Baker's result was later extended to deal with arbitrary Diophantine equations in two variables.

The difficulty in solving these particular cases suggested that the answer to Hilbert's problem was negative, and therefore that a general decision algorithm did not exist. The proof of this fact was established by Martin Davis, Hilary Putnam, Julia Robinson, and Yuri Matyasevitch. In 1960, the first three of them showed how to translate the halting problem into the language of Diophantine equations enriched by the addition of the exponential function (the behavior of any given program is described by an equation, in such a way that the program halts if and only if the equation has a solution), and in 1970 Matyasevitch eliminated the need for the exponential function.

A refinement of Matyasevitch's result shows that the case of Diophantine equations in nine variables is already undecidable, but it is not known whether this is the best possible result. In fact, Alan Baker conjectured that already in three variables the question would be undecidable.

4.2. Artificial Intelligence: Shannon's Analysis of the Game of Chess (1950)

None of the applications of the computer is more original and controversial than its use in artificial intelligence to simulate processes and results that are typical of human intelligence. The originality clearly stems from the intellectual provocation of considering thought—the most specifically human of all qualities—as something that machines can actually possess. The controversy results from the fact that artificial intelligence, especially in its earlier period during the 1950s and 1960s, compromised itself by making predictions that turned out to be, in practical terms, exaggerated and unrealistic, when not simply ridiculous.

In his famous 1950 article "Computing Machines and Intelligence," Turing himself had already suggested the possibility that machines could think. In particular, he proposed a practical criterion known as *Turing's test*: a machine may be said to think if a person exchanging written messages with it does not realize that the answers are not being given by another human being.

The term "artificial intelligence" was instead officially adopted by the computing community in 1956, during the historical congress held at Darmouth College in Hanover, New Hampshire. Among the participants were those who would become the most representative of the discipline, and who would later receive the highest recognition in computing, the *Turing Award*: Marvin Minsky in 1969, John McCarthy in 1971, and Allen Newell and Herbert Simon in 1975.

Artificial intelligence's original dream, explicitly formulated by Simon in the 1950s, was to succeed, within ten years, in writing programs capable of beating world chess champions, proving new and important mathematical theorems, and inspiring most theories in psychology.

Forty years later, most of the dream has been abandoned, and the computer's role was drastically downgraded. As a mathematical tool, the machine is nowadays almost exclusively used to perform massive calculations rather than to state and prove by itself new theorems, and as a model of the brain it has already been outdone by artificial neural networks. This does not mean that computers have not helped mathematicians to obtain some deep results and useful applications. The most significant examples, besides those we shall discuss below, are *expert systems*, which encode certain specialized knowledge into databases and make inferences through programming languages that mimic certain mechanical aspects of human reasoning.

In one domain only have Simon's predictions been fully realized, even if in a longer timespan than he had foretold: the game of chess. Already in 1864 Charles Babbage, the visionary inventor of the first computer, had envisaged the possibility of having a machine play chess, and he had even proposed a possible set of basic instructions. And in 1890 Leonardo Torres y Quevedo had completely formalized a strategy for checkmate when only the two kings and one rook are left.

But the first real analysis of the game in computing terms is due to an epoch-making article by Claude Shannon in 1950. In particular, he drew a clear distiction between several types of programs: (a) brute force *local* programs, which analyze the possibility tree up to a certain predetermined depth and choose the best move based on a minimax valuation, considering only the most promising moves (each level of depth can improve the program's performance by approximately two hundred Elo points); (b) *global* programs, which combine move depth analysis with an estimation of the distribution, mobility, balance, influence and control of the pieces; and (c) *strategic* programs, based on abstract rules similar to those used by humans.

The first game opposing a man to a program was played in 1951 between the computer scientist Alick Glennie and the Turochamp written by Alan Turing. Since the computers of the time were not powerful enough, Turing had to simulate the execution of the program by hand. And since the program was not very sophisticated, the game was easily won by Glennie in twenty-nine moves.

Simon's rosy predictions were shared by Mikhail Botvinnik, the world chess champion (except for two short periods) from 1948 to 1963, who in 1958 declared himself convinced that one day the computer would prove superior to man at the game. Botvinnik then undertook the development of global and strategic programs, which he pursued for many years.

Turing's test restricted to chess was successfully passed for the first time in 1980 by Belle, the world program champion (the first world tournament had been held in 1974). On one occasion, while the grandmaster Helmut Pfleger played twenty-six simultaneous games, three of these were secretly being played by the program. Five of the games, including one played (and won) by Belle, were then chosen and presented to various experts for analysis. One of these was the grandmaster Korchnoi, who had been the challenger for the world title in 1978. The majority of the experts, including Korchnoi and Pfleger but excluding Kasparov, failed to identify the game played by the computer.

The progress in chess playing programs has been truly impressive. In 1978, an international master was defeated for the first time when David Levy lost to Chess 4.7. Ten years later a grandmaster, Bent Larsen, was beaten by Deep Thought. And in 1996 it was a world champion's turn, Gary Kasparov's, to lose to a program—Deep Blue. Concurrently, in 1983 a program (Belle) became master for the first time, and another (Deep Thought) grandmaster in 1990. The final stage in this progression took place on 11 May 1997 when Deep Blue defeated world champion Kasparov, not just in one game but in a real tournament, by 3.5 to 2.5 points.

The programs up to Belle were of the local type, while Deep Thought and Deep Blue are global, but the design of strategic programs has so far proved impossible. This fact illustrates the philosophical limits of artificial intelligence's project, even in the domain were it was most successful: it has occasionally succeeded to simulate human thought by reproducing its outcome, but it has never been able to emulate it by reproducing its processes.

4.3. Chaos Theory: Lorenz's Strange Attractor (1963)

The fundamental problem of dynamics is to go from an implicit description of the laws governing the motion of a mathematical point or a physical body, to an explicit description of the trajectory followed by the point or the body; in a nutshell, to solve the equations of motion.

Classical dynamics has focused on motions described by linear differential equations, for which many analytical solution methods have been developed. On the other hand, the difficulty in solving nonlinear differential equations has prevented a comprehensive study of the phenomena that such equations describe, also due to the *instability* effect that these systems exhibit. Indeed, although in theory perfectly deterministic, nonlinear systems often behave in a virtually chaotic way, for small variations in the initial conditions may result in large variations in their solutions.

The advent of the computer has allowed mathematicians to tackle the study of nonlinear systems by the brute force afforded by computation: instead of solving the equations analytically, the process that these equations describe is simulated in an analog manner. The end result is not the equation of the trajectory but rather its image. Such a graphical solution often turns out to be good enough for the applications, as well as in a form that can be immediately seized by the eye of the imagination.

A classification of dynamical systems based on their behavior makes use of the notion of *attractor*, the stable configuration toward which the moving body tends. In the simplest case this attractor is a single point—a gravitational mass that attracts a body, for instance (thus the name attractor). Slightly more complex is the case of a closed curve, for example the ellipse described by the Earth rotating around the Sun.

Still more complex is the case of a surface that the body sweeps during a quasi-periodical motion obtained by the superposition of periodical motions—for instance, the motion of the Moon rotating around the Earth which is itself rotating around the Sun. The resulting surface is the composition of two perpendicular elliptical motions, and hence a kind of torus.

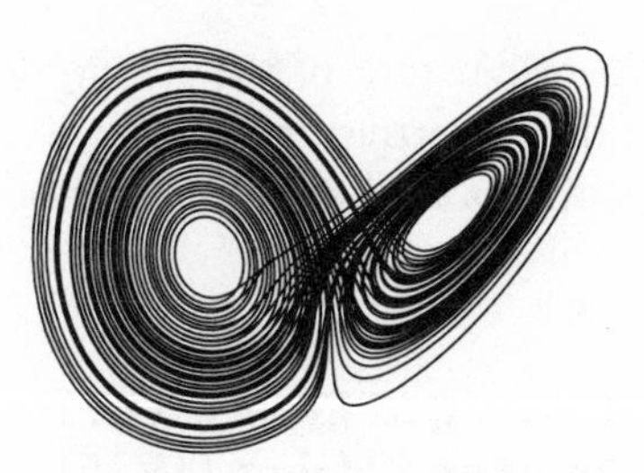

Fig. 4.2 The Lorenz attractor.

There are also *strange attractors* which, unlike the ones already mentioned, are nonclassical. Their strangeness consists in the fact that, instead of being points, curves, or ordinary surfaces, they are fractals (in a precise sense that will be defined below).

The first example of a strange attractor was found by Edward Lorenz in 1963, as the solution of the equations describing weather patterns he had proposed. This example became the symbol of the theory of chaos (fig. 4.2). An interesting feature of it is the fact that, precisely because such a solution is obtained by computer simulation, the general form of the Lorenz attractor is always more or less the same, but its details vary depending on the computer program used.

Only in 1995 did Konstantin Mischaikow and Marian Mrozek prove (ironically, by an extensive use of the computer) that the Lorenz system is really chaotic, in the sense that its behavior leads to a strange attractor. And in 2000 Warwick Tucker has proved that this attractor has exactly the shape that its computer-generated approximations show. This is not immediately obvious precisely because we are dealing with a chaotic system, for which small variations may produce large changes.

Besides the obvious interest generated by its practical applications, ranging from aerodynamics to meteorology, the computer simulation of nonlinear systems also raises interesting theoretical questions regarding the interpretation of its results. The chaos displayed on the computer screen is not automatic proof of the chaotic nature of the system described by the equations; and the real attractor of a chaotic system does not necessarily have the form of its approximations shown by the machine.

4.4. Computer-Assisted Proofs: The Four-Color Theorem of Appel and Haken (1976)

In 1852 Francis Guthrie was coloring a map of England when it occured to him that four colors appeared to be all one needed to color any map. To be sure, adjacent regions have to be painted in different colors, and the boundaries between two countries should not be too simple nor too complex. Not too simple means, for example, that the boundary cannot be a single point. Otherwise, by considering regions disposed as the slices of a pie we would conclude that no finite number of colors would be sufficient. Not too complex means that we have to exclude excessively jagged boundaries, for regions that all share the same boundary (the so-called Wada lakes, fig. 4.3) would necessitate as many colors as there are regions, whose number can be arbitrarily large.[1]

[1] Finding *two* regions with a common boundary is trivial: simply divide the plane into two parts with a straight line or a circle. Finding *three* (or more) regions with the same boundary is difficult, and requires a limiting process. To visualize it, suppose we had two lakes, one green and the other blue, both situated on a black island surrounded by a red sea. First we build a channel to bring red water onto the island in such a way that the distance from black earth to red water never exceeds one meter. Then we build another channel to bring green water onto the island so that the distance from black earth to green water never exceeds one-half meter. Finally we build a third channel, this time to bring blue water onto the island, in such a way that the distance from black earth to blue water never exceeds one-quarter meter. We now start refining the channels: the first channel is refined so that black earth is never more than one-eighth meter from red water, and so on. In the limit, the three regions, green, blue, and red, will be separated by a single black boundary of infinitesimally small width—that is, a black curve.

To prove that under the above restrictions at least four colors are necessary, it is enough to consider the configuration in figure 4.4, where each of the four countries shares a border with the other three. Augustus de Morgan showed that it is not possible for five countries to each share a border with the other four. But this only means that we cannot conclude, *using the same argument*, that five colors are necessary. Nor does it follow that four colors suffice, as believed by the many amateur mathematicians who during one century proposed many flawed proofs of the four-color conjecture.

In 1879 Alfred Kempe published a proof of the theorem, but in 1890 Percy Heawood discovered a mistake and observed that Kempe's argument only showed that *five* colors are enough. The proof consisted in showing (a) that regions sharing a boundary with at most five other regions (fig. 4.5) are unavoidable, in the sense that every normal map (i.e., such that there is no point at which more than three countries meet) must contain at least one such configuration; and (b) that maps containing unavoidable regions are reducible to other maps with fewer regions and can be painted with the same number of colors as the latter.

For instance, if a region is at most quadrilateral (i.e., if it shares a border with at most four other regions), the new map is obtained by contracting the region to a point (fig. 4.6). If the map so obtained can be colored with five colors or less, then the same is true of the original map: it is enough to use, for the region that became a point, a color that is different from the ones (at most four) used to paint the bordering regions.

A little more complicated is the case of pentagonal regions, or regions sharing a boundary with five others. The new map is then obtained by considering as a single region the portion of the map consisting of the pentagonal region together with two regions adjacent to it but not to each other (fig. 4.7). If the new map can be painted with at most five colors, no extra colors are needed to paint the original map: simply use, for the pentag-

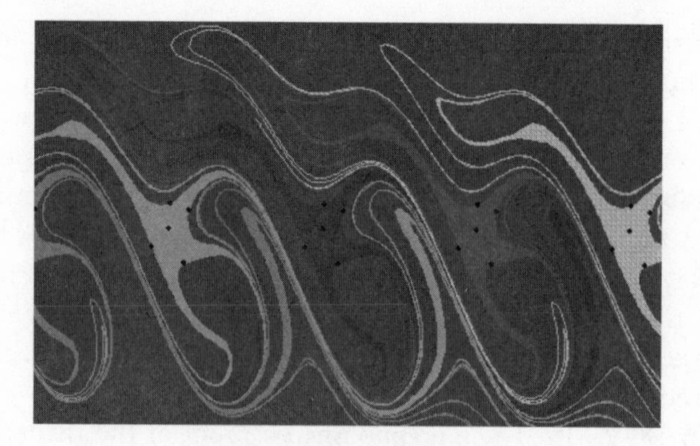

Fig. 4.3 Wada lakes.

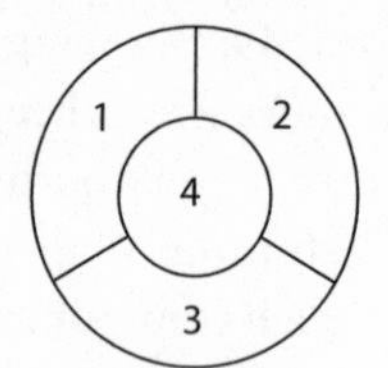

Fig. 4.4

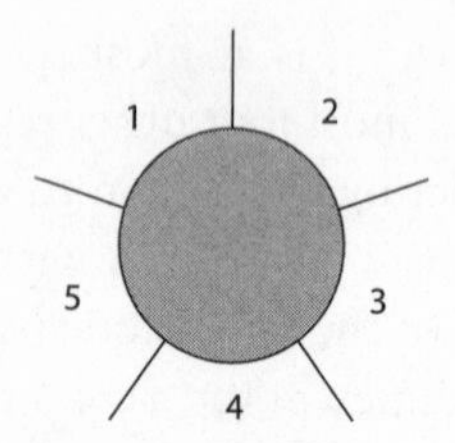

Fig. 4.5 Pentagonal region.

onal region, a color not already used to paint the four remaining regions in the new map (the two regions that have become one will be of the same color, but they will be separated by the pentagonal region, which will be of a different color).

In the four-color case, one can deal in a similar fashion with the regions that are at most triangular, and a trick allows one

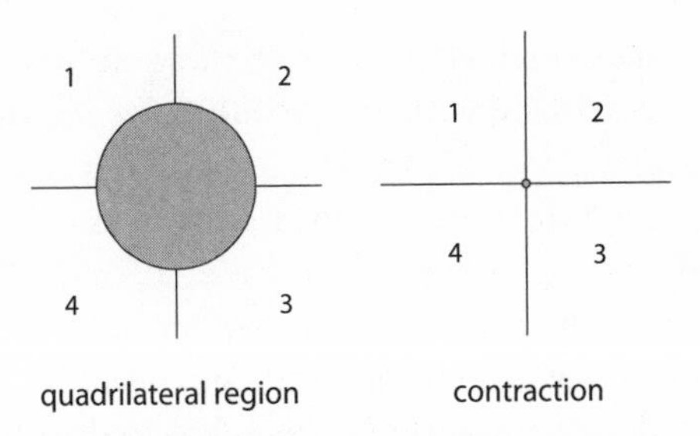

Fig. 4.6 Reducing a quadrilateral region.

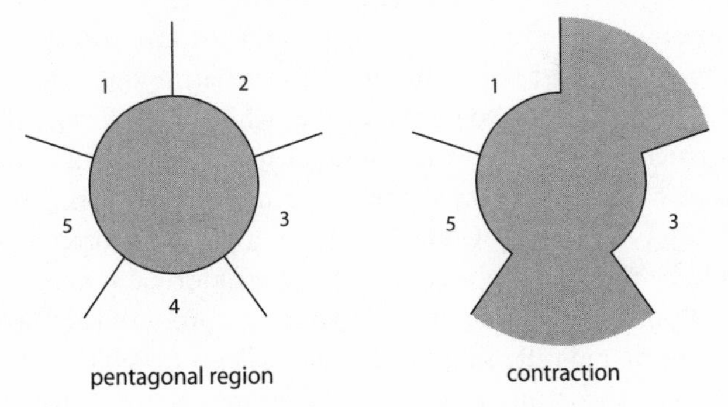

Fig. 4.7 Reducing a pentagonal region.

to dispose of the quadrilateral regions, but there is no way of treating the case of the pentagonal regions. All attempts to patch up Kempe's proof led on the one hand to increasingly large sets of unavoidable configurations and, on the other, to ever larger sets of reducible ones. Such arguments established the truth of the four-color conjecture for maps with up to one hundred regions. But it was only in 1976 that Kenneth Appel and Wolfgang Haken found a set of configurations that were at the same time unavoidable and reducible, thus proving the theorem in all its generality.

The interesting aspect of Appel and Haken's proof was not so much the solution of the problem, whose mathematical interest was rather limited, but the method they employed. The

1,482 unavoidable and reducible configurations were found by trial and error, starting with a set of 500 of them, through an interactive search process led by the computer, requiring 1,200 hours of machine time (equivalent to fifty days of continuous computing).

Hence, for the first time the proof of a mathematical theorem relied on computations that could not be verified by hand. When the paper with the proof was submitted to the *Illinois Journal of Mathematics*, the result was checked using a different program run on a different computer. This raises some philosophical questions, since computer-assisted proofs are *not* of the same kind as ordinary ones. In the latter, one goes directly from intuition to formalization, while in the former the computer program acts as an intermediary. The problem is that not only can't we know if the program correctly formalizes intuition, but that by Gödel's theorem a proof of correctness would be problematic, just as in the case of formal systems.

It is possible that one day this particular proof of the theorem will be radically simplified. But it is also possible for the four-color theorem to be instead symptomatic of a malaise common to all undecidable formal systems: the fact that there should exist theorems whose statements are short but whose proofs are arbitrarily long, for example, theorems of length n whose shortest proof has length at least 2^n. Otherwise, the system would be decidable, since to determine whether a proposition of length n is a theorem or not, it would be sufficient to generate all proofs of length at most 2^n and check if one of them proves the given proposition.

It is therefore not surprising if, on the one hand, simple propositions require complex proofs and, on the other, thousands of years of mathematical progress have probably exhausted the set of short (and interesting) proofs. What we are witnessing is perhaps the dawn of a new era, in which proofs will become increasingly long and complex. To cope with this problem there seems to be no other way out than to divide the work among many mathematicians, as in the case of the classification of finite groups, or to assign part of the task to the computer, as in the proof of the four-color theorem.

The most famous computer-assisted proofs of the century are those of the four-color theorem and of Kepler's conjecture, which we have already discussed. Another mathematically relevant example is the refutation of the Mertens conjecture, which went as follows.

In 1832 Möbius had considered those numbers in whose prime decomposition all factors have exponent equal to 1 (i.e., each prime factor occurs only once). He had assigned to such numbers the value 1 or −1 depending on whether the number of factors is even or odd, and he had defined the function $M(n)$ as the sum of all these values for numbers less than or equal to n. In 1897 Franz Mertens calculated the first 10,000 values of the function M and he conjectured that, for all n

$$-\sqrt{n} \leq M(n) \leq \sqrt{n}.$$

This question may appear to be of little interest but in fact Mertens's conjecture would have entailed the Riemann hypothesis, which is, as we shall see, the most important open problem in modern mathematics. The computation of greater and greater values of the function M seemed to confirm Mertens's intuition, but his conjecture was refuted in 1983 by Herman te Riele and Andrew Odlyzko, whose computer-assisted proof made extensive use of a CRAY supercomputer.

4.5. Fractals: The Mandelbrot Set (1980)

In 1906 Helge von Koch discovered that it is possible for a plane region to have a finite area but an infinite boundary. Consider an equilateral triangle and divide each side into three equal parts. Now take the central third of each side to be the

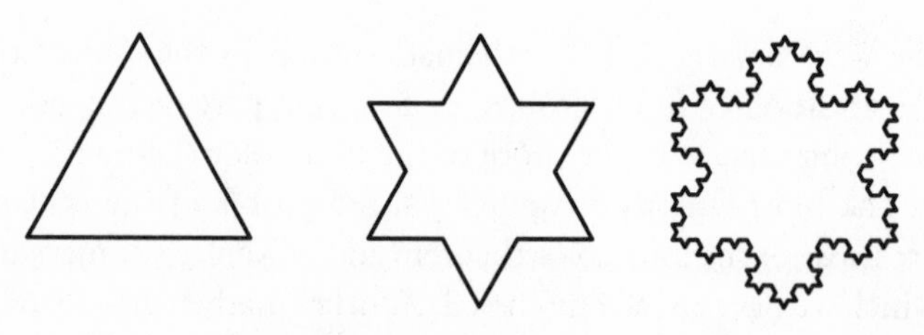

Fig. 4.8 Koch curve.

base of a new equilateral triangle, and repeat this process infinitely many times (fig. 4.8). The final product is a figure in the form of a snowflake, whose area is finite but with an infinitely long boundary (at each stage the length of the boundary increases by a factor of 4/3).

Due to the symmetric iteration of the process that defines it, the boundary of the Koch snowflake possesses the property of self-similarity. If any two segments in the various stages of the process are transformed—for example, one side of the starting triangle and one side of the little triangles in the first stage—the same limiting curve will result, only on a different scale.

Since curves such as Koch's cannot be measured in the usual way due to their infinite length, in 1918 Felix Hausdorff proposed to measure their degree of self-similarity by extending the notion of dimension in the following way. A segment is a one-dimensional self-similar figure that may be obtained by putting together two parts of magnitude one-half. Likewise, a square is a two-dimensional self-similar figure that may be obtained by putting together four parts of magnitude one-half. And a cube is a three-dimensional self-similar figure that may be obtained by putting together eight parts of magnitude one-half (fig. 4.9). In general, a d-dimensional self-similar figure would be one that may be obtained by putting together n^d parts of magnitude $1/n$. A Koch curve may be obtained putting together 4 parts of magnitude $1/3$ (a segment is di-

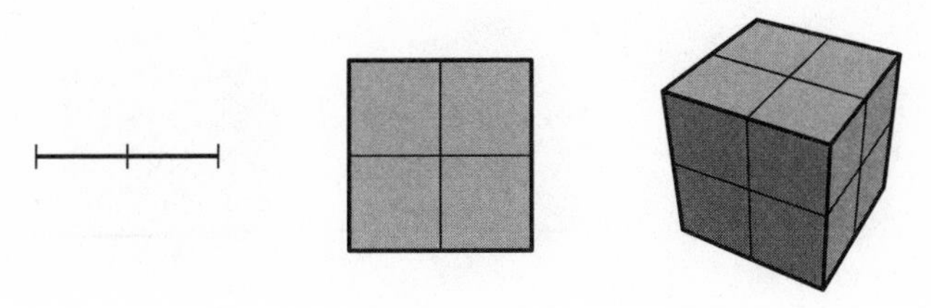

Fig. 4.9 Self-similar figures.

vided into three parts, and the central part is replaced by two equal parts), hence its dimension d is such that $4 = 3^d$, that is,

$$d = \frac{\log 4}{\log 3} \approx 1.26.$$

Figures having fractional dimension in the sense defined above are called *fractals*, and they exist in large numbers. For example, for each real number r between 1 and 2 there is a fractal curve of dimension r. Similarly, there are fractal surfaces with dimensions between 2 and 3. One example of these, called the *Menger sponge*, may be obtained as follows. Start with a cube and divide it into twenty-seven smaller cubes. Now remove the central seven of these smaller cubes (six on the faces and an internal one). Repeat this process infinitely many times (fig. 4.10). The dimension of the resulting surface is (approximately) 2.72, while the volume that it encloses is 0.

The above examples of fractals are highly regular and apply the same construction process at each stage. For this reason, if we enlarge a small portion, the resulting image is of the same type as the whole object. There are other fractals whose construction involves different processes at each stage, so that enlarging a detail will produce an image of a type different from that of the entire figure.

The study of this second kind of fractal began in the 1920s with Gaston Julia and Pierre Fatou, but it did not proceed very

Fig. 4.10 The Menger sponge.

far due to the demanding computations, which render the drawing of images by hand very difficult. The advent of the computer made it possible to revive the subject, and computer-generated images of elaborate fractals have now become a true and proper form of modern art.

The simplest type of fractal that may be considered, besides the one based on linear modifications of the initial figure, involves quadratic processes. In 1980 Benoit Mandelbrot discovered a kind of universal fractal that is defined in a rather indirect way, that is, using the transformation $x^2 + c$ of points in the plane (the values of x are therefore complex numbers, not just real numbers), and applying it repeatedly, starting at an arbitrary point.

If $c = 0$ there are three different cases: the points at a distance equal to 1 from the origin—that is, those on the unit circle—are invariant under the transformation (since x^2 equals x when x equals 1); the points inside the unit circle, whose distance from the origin is less than 1, move toward the origin (since x^2 is less than x when x is less than 1); the points outside the unit circle, at a distance greater than 1 from the origin, move toward infinity (since x^2 is greater than x when x is greater than 1). Thus, there are two regions of attraction, toward zero and toward infinity, separated by a circular boundary.

For an arbitrary c several things may occur: the number of regions of attraction may vary; besides the regions of attraction there may be regions of periodic orbits; and the boundary between the various regions is a fractal curve that may consist of

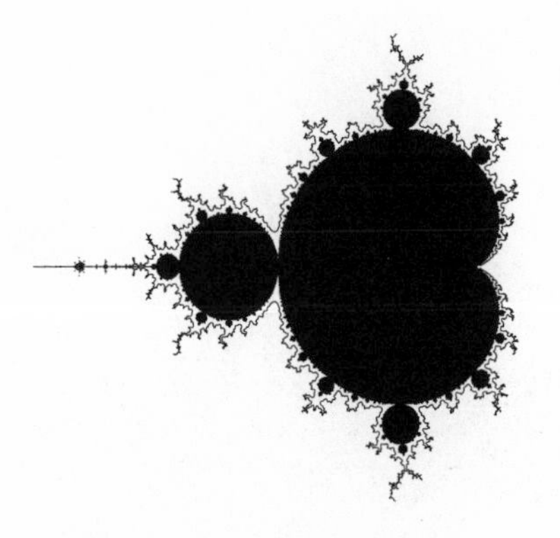

Fig. 4.11 The Mandelbrot set.

a single piece, of several pieces, or of only a myriad of scattered points.

The Mandelbrot set consists of the points c for which the boundary region is made up of one piece, and its strange appearance has become one of the most famous geometric forms (fig. 4.11). As Adrien Douady and John Hubbard have shown in 1985, the set itself is made up of one piece (in technical terms, it is a *connected* set), and as Jean Christophe Yoccoz has proved, each of its points that is not on the boundary is completely surrounded by a portion of the set made up of a single piece (i.e., the set is *locally connected*), one of the results for which Yoccoz was awarded the Fields Medal in 1994.

The position of a point c relative to the Mandelbrot set determines the behavior of the quadratic transformation $x^2 + c$. The importance of the study of this particular topic is emphasized by the award of the 1998 Fields Medal to Curtis McCullen, who isolated the points corresponding to transformations that define hyperbolic dynamical systems (that is, all of whose periodic orbits are circular), which are particularly useful and well known.

Despite its very particular definition, the interest in the Mandelbrot set is quite general. It is actually a system of refer-

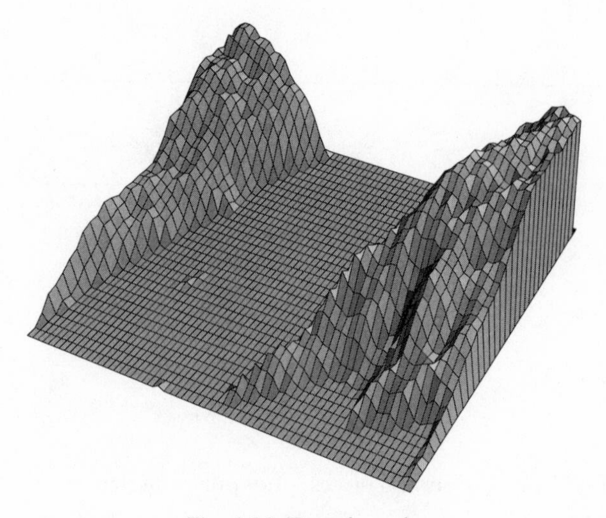

Fig. 4.12 Fractal graph.

ence for a complete study of complex dynamical systems, for it provides information not only on quadratic transformations, but also on any transformation that behaves as a quadratic one on some portion of the plane.

As for their applications, fractals are used to model objects that exhibit a structure on multiple scale levels, from maritime coastlines to mountain chains, and they are also used in computer graphics to render realistic images of such objects (fig. 4.12). Due precisely to the rich variety of the applications of fractals, Mandelbrot received the Wolf Prize in 1993, not for mathematics but for physics.

CHAPTER 5 OPEN PROBLEMS

Mathematics, as we hope to have shown, is essentially an activity of production and solution of problems: problems that can be easy or difficult, superficial or deep, theoretical or practical, pure or applied. Their supply is inexhaustible, also because the solutions are often the source of new problems. Having completed our treatment of the developments arising from Hilbert's problems, and more generally of twentieth-century mathematics, we may indulge to take a look at the problems of the future.

To be sure, it is hard to gauge the difficulty of a problem before having seen its solution, as Hilbert's problems have demonstrated. As examples of this, the third problem was solved by Max Dehn immediately after the Paris congress, and its solution appeared in print even before the publication of the congress proceedings. Similarly, the seventh problem was solved in 1929, despite the fact that ten years earlier Hilbert had considered a solution hardly possible before the end of the century.

Nevertheless, mathematicians consider that the problems they formulate are not only solvable but that they will be, sooner or later, actually solved. Here is how Hilbert put it in his Paris address: "The conviction that every problem has a solution is a powerful incentive for the researcher. In our hearts we hear the perpetual call: there is a problem, let us find the solution. And it can be found through the use of reason alone, because in mathematics there is no *ignorabimus*."

Hilbert wondered whether the solvability of every problem was peculiar to mathematical thought, or to a more general law of the nature of the mind. But he clearly stated that an acceptable solution to a problem could also consist in a proof of its unsolvability. This turned out to be the case for his first problem, on the continuum hypothesis, and for his tenth problem, on the existence of solutions of Diophantine equations.

Of course, negative solutions can be found throughout the entire history of mathematics. The irrationality of $\sqrt{2}$, discovered by the Pythagoreans, was nothing more than a proof of the impossibility of solving the equation $x^2 - 2 = 0$ in the rational numbers. In the nineteenth century, mathematicians showed that certain geometric or algebraic problems were impossible to solve—such as the squaring of the circle and the trisection of the angle using ruler and compass in the first case, and the solution by radicals of equations of degree greater than four in the second. But it was only in the twentieth century that the phenomenon attained a critical mass, also thanks to its clarification by means of Gödel's theorem.

It is with this warning—that a problem which appears interesting or solvable may turn out to be disappointing or unsolvable—that we shall propose a short list of open problems in mathematics, from what is probably the oldest one to one of the most recent ones, including the two problems that are universally considered the most profound: the Riemann hypothesis and the Poincaré conjecture.

5.1. Arithmetic: The Perfect Numbers Problem (300 B.C.)

Number theory is full of problems which, as Fermat's last theorem, are extremely easy to state but exceptionally hard to solve. The oldest open problem in mathematics is precisely of this kind.

In the sixth century B.C. the Pythagoreans had defined a *perfect number* as one that is equal to the sum of its divisors, excluding of course the number itself but including the number 1. For example, 6 and 28 are perfect numbers, their divisors being, respectively, 1 - 2 - 3 and 1 - 2 - 4 - 7 - 14. In *The Creation of the World* (III) the first-century Hebrew philosopher Philo Judaeus claimed that God created the world in six days precisely because 6 is a perfect number, and in *The City of God* (XI, 30) Augustine agreed.

Besides 6 and 28, the Greeks also knew that 496 and 8,128 were perfect. The fifth perfect number, 33,550,336, appears for the first time in a fifthteen-century German manuscript, and today only about forty perfect numbers are known. Around 300 B.C. Euclid proved that if $2^{n+1} - 1$ is prime, then $2^n(2^{n+1} - 1)$ is a perfect number (proposition IX.36 of the *Elements*), as can be easily verified. But it is not as easy to show that all the even perfect numbers are exactly those of the type found by Euclid. The proof of this fact was provided by Euler in 1737, using the same approach that allowed him to show that there are infinitely many prime numbers, and which led to the developments in relation with the Riemann hypothesis that we shall present later.

The even perfect numbers are therefore closely related to the prime numbers of the form $2^m - 1$, known as *Mersenne primes.* Euler discovered an efficient method to check whether $2^m - 1$ is prime or not, based on the so-called *little Fermat theorem:* if p is prime, then 2^{p-1} is equal to the identity of the cyclic group with p elements (or, equivalently, it is congruent to 1 modulo p).

Faithful to his habit, Fermat had only stated his little theorem, so Euler was forced to prove it. He gave a first proof in 1737, but in 1750 revisited the topic. For his second proof Euler developed the *theory of congruences*, or the theory of cyclic groups with a prime number of elements, which was to become one of number theory's most fertile tools.

Euler's test is nowadays used in the computer search for large prime numbers, and at the end of the twentieth century the largest known (Mersenne) prime was the previously mentioned $2^{6,972,593} - 1$, from which the largest known perfect number can be obtained.

As in the case of the famous Fermat theorem, the study of perfect numbers has also contributed to the development of vital segments of modern number theory. But a first problem remains open: the existence of odd perfect numbers.

If the answer is positive, an example could theoretically be found through an exhaustive search—for example, on a computer. In practice, however, it all depends on the size of the smallest odd perfect number. If the answer is negative, then Euclid's and Euler's results provide a complete characterization of perfect numbers.

Whatever the case, there is a second open problem: whether the set of even perfect primes is finite or infinite. Or, equivalently, whether there are finitely or infinitely many Mersenne primes.

5.2. Complex Analysis: The Riemann Hypothesis (1859)

Every integer can be decomposed, with respect to addition, as the sum of terms all equal to 1. With respect to multiplication, on the other hand, there exist *prime numbers* that cannot be decomposed, i.e., they do not admit factors, or divisors, different from themselves and 1. The prime numbers are thus the atoms of the world of numbers, and their study may be compared to that of the elementary particles in the physical world.

The first significant results in this area were due to the Greeks, who showed that every number may be expressed in a unique way as a product of primes, and that the sequence of prime numbers is infinite, even though they become more and more rare in the sequence of all numbers.

A direct proof of the infinitude of the prime numbers appears in Euclid's *Elements* (IX, 20), but a surprinsigly indirect proof was given by Euler in 1737. He remarked that, since every number is a product of primes, as n varies so do all possible products of primes raised to all possible exponents. If there were only a finite number of primes, the sum

$$1 + \frac{1}{2} + \frac{1}{3} + \ldots + \frac{1}{n} + \ldots$$

would be finite, for it would be the product of a finite number of geometric progressions of the form

$$1 + \frac{1}{p} + \frac{1}{p^2} + \ldots = \frac{p}{p-1}.$$

But the former sum is infinite, because the two fractions $1/3$ and $1/4$ contribute at least $1/2$, and so do the next four, eight, sixteen, etc.

There are 25 prime numbers from 1 up to 100; 168 up to 1,000; 1,229 up to 10,000; 9,592 up to 100,000. A distribution which, as Euler and Gauss both observed, decreases in an approximately logarithmic fashion, in the sense that the number of prime numbers up to 10^n is about $10^n/2n$, e.g., 25 up to 100; 167 up to 1,000; 1,250 up to 10,000; 10,000 up to 100,000. In general, and using natural logarithms, we can state as a conjecture the *prime number theorem*: the number of prime numbers up to n approaches the ratio

$$\frac{n}{\log n}$$

as n tends to infinity.

In 1859, while trying to prove the conjecture, Bernhard Riemann observed that the problem is related to the behavior of the function

$$\zeta(z) = 1 + \frac{1}{2^z} + \frac{1}{3^z} + \cdots + \frac{1}{n^z} + \cdots.$$

The connection between the function ζ (known as Riemann zeta function) and the prime numbers is clear from the previous proof by Euler, who showed, however, that the value of ζ is infinity when the variable z is less than or equal to 1. For this reason Riemann extended the function ζ from real to complex numbers, through a technique known as analytic extension (basically, the value of ζ is defined not as the limit of the partial sums but as the limit of their mean).

The zeta function has an infinite number of nonreal complex zeros, that is, numbers of the form $z = x + iy$, with $y \neq$ and $\zeta(z) = 0$, and these are all located in the band of the complex plane defined by taking x between 0 and 1. Riemann conjectured that all the zeros of ζ lie on the line with equation $x = 1/2$. This is the so-called *Riemann hypothesis*, the most important open problem in modern mathematics. What we know so far is that infinitely many zeros of ζ do lie on that line, as Godfrey Hardy proved in 1914, and that such is also the case for the first few billions of zeros.

To prove the prime number theorem it was not necessary to know the property of th zeta function mentioned in the Riemann hypothesis. In fact, the theorem was proved in 1896 by Jacques Hadamard and Charles Jean de la Vallée Poussin, and their proof only required showing that no zero of the zeta function lies on the line with equation $x = 1$.

Thus, the Riemann hypothesis remained open, and it became part of *Hilbert's eighth problem.* This problem also asked several other questions about the prime numbers, such as the *Goldbach conjecture*, dating from 1742, and the *twin primes conjecture.* The first of these conjectures states that every even number greater than 2 is the sum of two primes, and the second one that there are infinitely many prime numbers whose difference is 2 (such as 3 and 5, or 10,006,427 and 10,006,429). Like the Riemann hypothesis, both these conjectures remain so far unproven.

Hilbert also proposed to study the behavior of the (ideal) prime numbers over an arbitrary field. A version of the Riemann hypothesis for an equivalent of the zeta function associated to algebraic curves over finite fields was proposed by Emil Artin in 1924 and proved in 1940–41 by André Weil, the 1979 Wolf Prize winner. In 1949, Weil proposed his own conjecture, a version of the Riemann hypothesis for multidimensional algebraic manifolds over finite fields, which became known as the *Weil conjecture.* It was proved in 1973 by Pierre Deligne, who was awarded the 1978 Fields Medal for this achievement. Deligne's proof was the first significant result obtained through the use of an arsenal of extremely abstract techniques in algebraic geometry (such as schemas and *l*-adic cohomology) introduced in the 1960s by Alexandre Grothendieck, the 1966 Fields medalist.

The apparent dissociation from the problems and techniques of classical number theory did not mean that traditional problems were completely abandoned. From Deligne's result, for instance, there follows a conjecture due to Ramanujan and dating back to the beginning of the century. Moreover, the methods employed by Deligne are the same ones that allowed Faltings and Wiles to prove the Mordell conjecture in 1983 and Fermat's theorem in 1995. After the arithmetic and analytic methods introduced by Fermat and Euler, with the infinite descent method and the introduction of the zeta function, respectively, the last quarter of a century has therefore seen the advent of a new approach to number theory through techniques of algebra and geometry.

Once a problem in number theory has been solved using analytic or algebraic-geometric techniques, we may ask whether these methods are really necessary, or if, on the contrary, it is possible to find classical proofs not based on concepts foreign to number theory itself. Such proofs are called "elementary," from the point of view of their logical complexity—which must not be confused with their mathematical complexity, since the use of more limited techniques usually results in more complex proofs.

In the case of the prime number theorem, Paul Erdös and Atle Selberg provided an elementary proof of the theorem in 1949, which earned Selberg the Fields Medal in 1950, and both of them the Wolf Prize, in 1983–84 and 1986, respectively. Elementary proofs of either Mordell, Ramanujan, or Fermat conjectures have not yet been found, and it is thought that such proofs may have an exorbitant length and complexity.

5.3. Algebraic Topology: The Poincaré Conjecture (1904)

Algebraic topology is the study of topological properties by means of algebraic methods. The first example of this approach is the *Euler characteristic* of a surface, already known to Descartes in 1639 and to Leibniz in 1675, but rediscovered and published by Euler in 1750.

The starting point is the observation that, given any convex polyhedron, the following relation holds among the number V of its vertices, E of its edges, and F of its faces:

$$V - E + F = 2.$$

For example, a cube has 8 vertices, 12 edges, and 6 faces, hence $8 - 12 + 6 = 2$.

The relation is still valid for any graph drawn on the surface of a sphere, which shows that we are in effect dealing with a topological property: if a rubber polyhedron is inflated in the shape of a sphere, its edges become a graph on the spherical surface and, conversely, if we flatten out the faces of a graph drawn on the surface of a rubber sphere we obtain a polyhedron.

What makes the situation interesting is the fact that the quantity $V - E + F$ depends only on the type of surface on which the graph is drawn. The value of this quantity is $2 - 2n$ if the surface is a sphere with n handles, and $2 - n$ if it is a sphere with n Möbius strips. For instance, the value is 2 for the sphere, 1 for the projective plane, and 0 for the torus and the Klein bottle. Hence, knowing whether a given closed, 2-dimensional surface is orientable or not, together with its Euler characteristic provides a complete characterization of the surface.

For surfaces of dimension 3 or higher, a concept analogous to the Euler characteristic was defined by Poincaré in a series of articles between 1895 and 1900, but it is not enough to classify these surfaces. The idea is then to review the previous results in more detail, and to associate to every 2-dimensional surface not just a single number but a *fundamental group*. After choosing some fixed point on the surface, one considers closed paths that start and end at that point (the composition of these paths consists in tracing out one path after the other; the identity path is the "path" that stays put, i.e., never leaves the point; and the inverse of a given path is the path that goes through the same points in the opposite direction).

Since we are dealing with topological properties, the paths must be seen as made out of rubber: two paths that can be transformed into one another by stretching or contracting them without breaking them are essentially the same. This identification of paths is called *homotopy*, and for this reason the fundamental group of a surface is also called the first homotopy group.

The fundamental group of the sphere is trivial, since any closed path on it can be contracted to a single point. Moreover, the sphere is the only closed and orientable surface whose fundamental group is trivial. For if a surface has at least one handle, a path around the base of the handle cannot be contracted to a point. The fundamental group is therefore enough to distinguish the sphere from any other orientable surface and, more generally, all the different 2-dimensional surfaces among themselves.

Poincaré extended the notion of a fundamental group to surfaces of dimension 3 and higher, and hoped that his generalization would lead to a topological classification of these surfaces by algebraic means. However, things turned out to be more complicated than expected, and we know now that the fundamental groups are not enough to classify all 3-dimensional surfaces. For this reason, Thurston's classification, that we have previously mentioned, makes an essential use of concepts that are not only algebraic but also geometric, such as the possible types of geometries that can be defined on the components of a surface.

In 1904, Poincaré formulated a conjecture regarding not arbitrary surfaces but only the hypersphere. He asked whether the hypersphere is the only closed and orientable tridimensional surface whose fundamental group is trivial. An affirmative answer would follow from Thurston's characterization of tridimensional surfaces, but this result has not yet been proved. What is more, the Poincaré conjecture is precisely one of the greatest obstacles to the completion of the proof.

The interesting thing is the fact that, if we extend the conjecture to spheres of arbitrary dimension, the only open case is the original one raised by Poincaré. For spheres of dimension 5 or higher, the Poincaré conjecture was in effect proved in 1960 by Stephen Smale, who was awarded the Fields Medal in

1966 for this result. (Subsequently, Smale became one of the most famous American intellectuals to denounce the U.S. involvement in the Vietnam War, and at one point the University of California stopped paying his salary.) As for the 4-dimensional sphere, the conjecture follows from Freedman's characterization of 4-dimensional surfaces, in a way similar to what was done for 2-dimensional surfaces.

Independently of the solutions, the difficulties in proving the Poincaré conjecture have shown that the information contained in the fundamental group is too restricted. For this reason, Witold Hurewicz introduced in 1935 an infinite sequence of *homotopy groups* for the n-dimensional sphere. The fundamental group is the first one in this sequence, and the first n are the *homology groups*, obtained by considering multidimensional, rather than unidimensional, paths—for example, not only elastic bands stretched out on the surface of the sphere but also small rubber balls that can be inflated or deflated, and so forth.

The fundamental result on the successive homotopy groups of the n-dimensional sphere is the *finiteness theorem*, which Jean-Pierre Serre proved in 1951: all these groups are finite, except the $(2n - 1)$th group for n even—for example, the third group of the 2-dimensional sphere. This result won Serre the Fields Medal in 1954, and it contributed to his winning of the Wolf Prize in 2000.

The precise determination of these homotopy groups turned out to be very complicated. The first two were calculated in 1950 by Lev Pontryagin, and the third one by Rokhlin in the same year, while Serre calculated the fourth group one year later. To carry out his calculation, Pontryagin had to determine under which conditions a compact n-dimensional surface is the boundary of some $n + 1$-dimensional surface. He found a necessary condition, which René Thom proved to be also sufficient in 1954.

From this latter result *cobordism theory* was born, and for it Thom obtained the Fields Medal in 1958. Two of the most spectacular applications of cobordism also led to Fields Medal awards in 1962 and 1966. These were John Milnor's theorem on exotic spheres, which in this context may be reformulated by saying that in dimension 7 there are spheres that are not the boundary of a ball, and the index theorem of Michael Atiyah and Isadore Singer. The extension, due to Milnor and Smale, of cobordism to *h*-cobordism (*h* for homotyopy) allowed Sergei Novikov to obtain the Fields Medal in 1970 for his classification of differentiable manifolds of dimensions greater than or equal to 5.

5.4. Complexity Theory: The P = NP Problem (1972)

Turing's definition of an algorithm separates the numerical functions into two classes: computable and noncomputable. This division is only a first approximation, for many functions that are computable in theory are not at all computable in practice. For instance, an algorithm whose execution would require an amount of time greater than the life of the universe, or simply greater than the duration of a human life, cannot be considered as actually executable, even if it is so in principle.

From the point of view of the applications, it is therefore necessary to consider only those algorithms whose execution is sufficiently fast. In 1965 Edmonds and Cobham proposed, as a second approximation, the distinction between algorithms that can be run in *polynomial time* and those that cannot. The execution time is measured by the number of steps performed by the computer, and the variable of the polynomial corresponds to the size of the data on which the algorithm operates—its length, for example. Thus, a quadratic algorithm does not require more than 100 computational steps on 10-digit numbers; not more than 10,000 steps on 100-digit numbers, and so on.

Of course, the execution time of an algorithm depends heavily on the type and power of the computer on which it is run. Perhaps surprisingly, it turns out that if a given algorithm operates in polynomial time on some particular machine, it also operates in polynomial time when run on any other computer. In other words, the difference between various computer models and their different implementations affects the execution time only by a polynomial factor, which combined with a polynomial time execution does not change the (polynomial time) nature of the algorithm. To be executable in polynomial time is thus an intrinsic, and not a contingent, characteristic of an algorithm.

Among the algorithms we have presented so far, the simplex method is nonpolynomial, given that for an infinity of data the algorithm requires an exponential time to provide an answer. This does not mean that the linear programming problem cannot be solved in polynomial time, but only that the particular solution provided by the simplex method is not a polynomial time solution. And, in fact, in 1979 Khachian found an alternative algorithm, the *ellipsoid method*, that solves the linear programming problem in polynomial time.

The class of problems for which a polynomial time solution exists is denoted by *P*. In 1972 Stephen Cook, Richard Karp, and Leonid Levin discovered a class, potentially larger than *P*, that is known as *NP*. The problems in *NP*, although not necessarily solvable in polynomial time, have "almost" that property, in the sense that for any candidate solution, one can verify in polynomial time whether it is indeed a solution or not. The difference between *P* and *NP* is therefore the following: for a problem to belong to the first class it is necessary that there exist a method to *find* the solution in polynomial time, while to belong to the second class it is enough that a method to *verify* the solution of the problem in polynomial time exists.

It is easy to be convinced that finding a solution is more difficult than checking it out. For example, to verify that a given telephone number is actually the number of a certain person is easy, for it is sufficient to look up the person's name and number in the telephone book. But to find the person whose number is a given number is difficult, for it requires the exhaustive search through the entire telephone book.

Here is a more mathematical illustration. To verify that

$$4{,}294{,}967{,}297 = 641 \times 6{,}700{,}417$$

is child's play, but to find the factorization on the right requires the genius of Euler or the power of a computer. The problem of factor decomposition is precisely one of those in *NP*, because it is easy to verify if two given numbers are or are not the factors of a third one. But it is not known whether the problem is also in *P*, that is, if there is a faster method to test whether a number is composite or prime (it is known that such a method exists if the Riemann hypothesis is true).

This latter fact is precisely the basis for ***private key cryptography***, whose general idea is the following. The sender and the receiver each possess a very large integer, which acts as a personal (and secret) coding and decoding key. To send a message m to the receiver, the sender encodes the message using his or her own key c, thus transforming it into mc. The receiver encodes the incoming message mc using his or her own key d, transforms it into mcd, and then sends it back to the sender. The latter decodes mcd using c, and sends back md to the receiver, who, finally, using the key d recovers the original message m. The efficiency of the method rests on the fact that the double decoding of the message requires finding the factors of very large numbers, a task that can be rapidly accomplished only by knowing the keys. The drawback of the method is that it requires a double encoding and decoding by both parties, the sender and the receiver.

This obstacle can be circumvented by the use of *public key cryptography*, based on a similar but more complicated idea. Each correspondent in the network possesses two very large integers that act as keys: one is c (the coding key), which is made public, and the other is d (for decoding), which is kept secret. To send a message m to the receiver, the sender encodes it using the public key c to transform it into m^c. The receiver then decodes this latter message using his or her secret key d, transforming it into $(m^c)^d = m^{cd}$. To successfully decode the message, the final message must be the same as the original one, that is, cd must be equal to 1. Although this is literally impossible, Fermat's little theorem guarantees that, given numbers p and q, if cd equals 1 modulo $(p - 1)(q - 1)$, then m^{cd} equals m modulo pq. The efficiency of the method stems from the fact that to encode and decode the message it is enough to know the product pq, which is also made public, but finding the decoding key d from the encoding key c requires the knowledge of $(p - 1)(q - 1)$, a number that can be easily obtained from decomposing pq, but such a decomposition cannot be rapidly performed.

In general, thousands of problems of theoretical or practical interest are known to belong to the class *NP*, without knowing if they are also in *P*. Examples of such problems related to questions we have discussed are the satisfiability of propositional formulas, the existence of integer solutions of quadratic Diophantine equations, and the possibility of coloring a map with three colors. An example of a problem in the calculus of variations is the *Steiner problem*, for which it is possible in certain cases to obtain an empirical solution using soap bubbles: given a geographical map, connect the cities with roads so that the total length of the road network is as small as possible (the solution obtained using soap bubbles is locally, but not always globally, optimal). A similar and very famous example, due to its numerous practical applications, is the *traveling salesman problem*: given a map with cities connected by roads, find a path of minimal length that visits each city exactly once.

One of the surprising discoveries made by Stephen Cook, Richard Karp, and Leonid Levin was that all these problems (with the only possible exception of the factorization problem), as well as thousands of others in the most diverse branches of pure and applied mathematics, are essentially equivalent, in the sense that finding a polynomial-time solution for any one of them would also mean finding polynomial-time solutions for all. This is due to the fact that there are polynomial-time "translations" between any two of them. For this result Cook and Karp received the Turing Award, in 1982 and 1985, respectively. Levin, on the other hand, ended up in jail as a dissident, and after his release, thanks to Andrej Kolmogorov's intercession, emigrated from the Soviet Union.

Finding a polynomial-time solution, or showing that none exists, for any one of the equivalent problems identified by Cook, Karp, and Levin has so far proved an impossible task. The question of whether *P* and *NP* are one and the same class or not has therefore become a challenge, and may be considered the most famous open problem in theoretical computer science.

To put the problem in purely mathematical terms, let us recall that the famous 1890 *Nullstellensatz* of Hilbert provided a necessary and sufficient condition for a finite system of polynomial equations with complex coefficients to have a solution. Dale Brownawell showed in 1987 that the problem can be solved in exponential time, but it is not known whether it could also be solved in polynomial time. If the coefficients and the solutions of the equations are restricted to the rational numbers (or even to the numbers 0 and 1), a polynomial-time solution of the problem exists if and only if *P* equals *NP*. Thus, our story closes, appropriately, under the sign of Hilbert's vital spirit—the same spirit that has pervaded it.

CONCLUSION

Having reached the end of our journey through twentieth-century mathematics, it is time to review its various stages. The asynchronous, collagelike nature of our presentation requires perhaps a complementary angle that would highlight the main threads in the fabric of the story. We list them below, in the form of chronological tables.

Problems and Conjectures

First of all, there are the problems and conjectures that have guided our story of the search for their solutions. Here are the most significant ones:

300 B.C.	Euclid	Perfect numbers
1611	Kepler	Maximum density configuration of spheres
1637	Fermat	Integer solutions of $x^n + y^n = z^n$
1640	Fermat	Prime numbers of the form $2^{2^n} + 1$
1742	Goldbach	Even integers as sums of two primes
1847	Plateau	Minimal surfaces
1852	Guthrie	Coloring of maps with four colors
1859	Riemann	Zeros of the zeta function
1883	Cantor	Continuum hypothesis

1897	Mertens	Bounds to the Möbius *M* function
1902	Burnside (I)	Finitely generated periodic groups
1904	Poincaré	Characterization of the hypersphere
1906	Burnside (II)	Simple groups of odd order
1922	Mordell	Diophantine equations with infinitely many rational solutions
1928	Hilbert	Decidability of first-order logic
1933	Robbins	Axiomatization of Boolean algebras
1949	Weil	Riemann hypothesis over finite fields
1955	Taniyama	Parameterization of elliptic curves
1962	Shafarevich	Reduction of equations modulo prime numbers
1972	Cook, Karp, and Levin	$P = NP$
1979	Conway and Norton	Moonlight

The problems proposed by Hilbert in 1900 have been one of the two leading threads of our story. Here, in a separate list, are those we have mentioned in the text:

First	Continuum hypothesis
Second	Consistency of analysis
Third	Decomposition of the tetrahedron
Fourth	Geodesics in various geometries
Fifth	Locally Euclidean groups and Lie groups
Sixth	Axiomatizations of probability and physics
Seventh	Transcendentality of e^π and $2^{\sqrt{2}}$
Eighth	Riemann hypothesis, Goldbach conjecture
Tenth	Solutions of Diophantine equations
Eighteenth	Crystallographical groups, Kepler's problem
Nineteenth	Analyticity of solutions of variational problems
Twentieth	Existence of solutions of variational problems
Twenty-third	Calculus of variations

Results

The other leading thread of our presentation has been the works of Field Medal and Wolf Prize winners. For the majority of them, we have tried to call attention to their most significant results. Of those who were awarded the Fields Medal we have recalled the following:

1936	Douglas	Plateau's problem
1950	Schwartz	Theory of distributions
1950	Selberg	Prime number theorem
1954	Kodaira	Classification of algebraic manifolds in 2 dimensions
1954	Serre	Homotopy groups of *n*-dimensional spheres
1958	Roth	Rational approximation of irrational algebraic numbers
1958	Thom	Cobordism theory
1962	Hörmander	Hypoelliptic operators
1962	Milnor	Exotic structure of the 7-dimensional sphere
1966	Atiyah	K-theory, index theorem
1966	Cohen	Independence of the continuum hypothesis
1966	Grothendieck	Schemas, *l*-adic cohomoly
1966	Smale	Poincaré conjecture in dimensions ≥ 5
1970	Baker	Extension of the Lindemann and Gelfond theorems
1970	Hironaka	Resolution of singularities of algebraic manifolds
1970	Novikov	Classification of differentiable manifolds of dimension ≥ 5
1970	Thompson	Second Burnside conjecture
1974	Bombieri	Number theory, minimal surfaces
1978	Deligne	Weil conjecture

1983	Connes	Algebras of von Neumann operators
1983	Thurston	Classification of 3-dimensional surfaces
1983	Yau	Calabi-Yau manifolds
1986	Donaldson	Exotic structure of 4-dimensional space
1986	Faltings	Shafarevich and Mordell conjectures
1986	Freedman	Classification of 4-dimensional manifolds
1990	Jones	Invariants of knots
1990	Witten	Superstring theory
1990	Mori	Minimal program for 3-dimensional algebraic varieties
1994	Bourgain	Hilbert subspaces of Banach spaces
1994	Yoccoz	KAM theorem, Mandelbrot set
1994	Zelmanov	Restricted first Burnside conjecture
1998	Borcherds	Moonlight conjecture
1998	Gowers	(Non)symmetrical Banach spaces
1998	Kontsevich	Invariants of knots
1998	McCullen	Mandelbrot set

Wolf Prize winners we have mentioned:

1978	Siegel
1979	Weil
1980	Kolmogorov
1982	Whitney
1983–84	Erdös
1984–85	Kodaira
1986	Eilenberg, Selberg
1988	Hörmander
1989	Milnor
1990	De Giorgi
1992	Thompson
1993	Mandelbrot (physics)
1994–95	Moser
1995–96	Langlands, Wiles
2000	Serre

Besides the works of mathematicians, we have also quoted, at least in passing, the results of some computer scientists who have received the highest recognition in their field, namely, the Turing Award:

1969	Minsky	Artificial intelligence
1971	McCarthy	Artificial intelligence
1975	Newell and Simon	Artificial intelligence
1976	Scott	Lambda calculus semantics
1982	Cook	Complexity theory
1985	Karp	Complexity theory

Finally, certain works in applied mathematics are directly connected to results that have won for their authors or for someone else the Nobel Prize in various fields:

1932	Heisenberg	Physics, quantum mechanics
1933	Schrödinger	Physics, quantum mechanics
1962	Crick and Watson	Medicine, DNA structure
1969	Gell-Mann	Physics, symmetry of quarks
1972	Arrow	Economics, social choice, general equilibrium
1975	Kantorovich and Koopmans	Economics, linear programming
1976	Prigogine	Chemistry, dissipative systems dynamics
1979	Glashow, Weinberg, and Salam	Physics, symmetry of electroweak forces
1983	Debreu	Economics, general equilibrium
1994	Nash	Economics, game theory

REFERENCES AND FURTHER READING

For General Readers:

Lang, Serge. *The Beauty of Doing Mathematics.* Springer-Verlag, 1985.

Dieudonné, Jean. *Mathematics—The Music of Reason.* Springer-Verlag, 1992.

Devlin, Keith. *Mathematics: The New Golden Age.* Columbia University Press, 1999.

Tannenbaum, Peter, and Robert Arnold. *Excursions in Modern Mathematics.* Prentice Hall, 1995.

Stewart, Ian. *From Here to Infinity: A Guide to Today's Mathematics.* Oxford University Press, 1996.

Casti, John. *Five Golden Rules: Great Theories of 20th-Century Mathematics, and Why They Matter.* Wiley, 1996.

———*Five More Golden Rules: Knots, Codes, Chaos, and Other Great Theories of 20th-Century Mathematics.* Wiley, 2000.

Mathematical Intelligencer, a magazine of mathematical popularization published every three months by Springer-Verlag, New York (175 Fifth Avenue, New York, NY 10010).

For Advanced Readers:

Kline, Morris. *Mathematical Thought from Ancient to Modern Times.* Oxford University Press, 1972.

Browder, Felix, ed. *Mathematical Developments Arising from Hilbert Problems.* American Mathematical Society, 1976.

Halmos, Paul. "Has Progress in Mathematics Slowed Down?" In *American Mathematical Monthly* 97 (1990): 561–588.

Casacuberta, Carles, and Manuel Castelet, eds. *Mathematical Research Today and Tomorrow: Viewpoints of Seven Fields Medalists.* Springer-Verlag, 1992.

Pier, Jean-Paul, ed. *The Development of Mathematics, 1900–1950.* Birkhäuser, 1994.

Kantor, Jean-Michel. "Hilbert's Problems and Their Sequel." In *Mathematical Intelligencer* 18 (1996): 21–30.

Monastyrsky, Michael. *Modern Mathematics in the Light of the Fields Medals.* AK Peters, 1997.

Atiyah, Michael, and Daniel Iagolnitzer, eds. *Fields Medalists' Lectures.* World Scientific, 1997.

Smale, Stephen. "Mathematical Problems for the Next Century." In *Mathematical Intelligencer* 20 (1998): 7–15.

Pier, Jean-Paul, ed. *The Development of Mathematics, 1950–2000.* Birkhäuser, 2000.

Arnol'd, Vladimir, Michael Atiyah, Peter Lax, and Barry Mazur, eds. *Mathematics Tomorrow.* International Mathematical Union, 2000.

INDEX